WINNING WAYS

Copyright © 2004 The Ogden Newspapers, Inc.
BookMaster, Inc.
2541 Ashland Road, P.O. Box 2139
Mansfield, OH 44509

All Rights Reserved

No part of this publication may be reproduced, stored in a retrieval system or transmitted in any form by means of electronic, mechanical, photocopying, recording or otherwise, without written permission of The Ogden Newspapers, Inc., 1500 Main Street, Wheeling, West Virginia 26003.

1st Printing 2004
Printed in the United States of America

Library of Congress Cataloging-in-Publication Data

Novotney, Steven
Winning Ways – Tales of Past Pittsburgh Pirates
cm.
ISBN 0-9727744-0-8
1. Pittsburgh Pirates – non-fiction. 2. baseball – non-fiction. 3. World Series. 4. Pittsburgh, Pa.

*For the heroes of my family,
from my immigrant grandmother who
dared to dream on American soil,
to my son who has protected the freedoms
she believed in without pause.*

Contents

Foreword VII
Acknowledgements XV

Chapter 1	Ralph Kiner 1
Chapter 2	Bob Friend 13
Chapter 3	ElRoy Face 25
Chapter 4	Bill Virdon 35
Chapter 5	Bill Mazeroski 49
Chapter 6	Steve Blass 61
Chapter 7	Manny Sanguillen 71
Chapter 8	Dave Guisti 83
Chapter 9	Nellie Briles 95
Chapter 10	Kent Tekulve 107
Chapter 11	Bill Robinson 123
Chapter 12	Chuck Tanner 135
Chapter 13	Barry Bonds 149
Chapter 14	Lloyd McClendon 165

Foreword

By Chuck Tanner
Pirates Manager, 1977-85

There are more than a million aspects that must come together for a big-league baseball team to win, and capturing a world title is the most difficult accomplishment of all. Pirates fans across the globe have been blessed with three championship teams in the last 50 years, and I am very proud to have played a part with one of those special ballclubs.

When I am asked about that 1979 World Series team, what still sticks out the most in my mind is very simple – I had the best collection of players and they formed the best team. Sure, Cincinnati and Baltimore had more superstars than we did, but we won because we had a group of players that ultimately showed the world how to win.

From beginning to end, I'll never forget the Series with Baltimore in '79, but the story of that season starts before we played our first game. We didn't have a championship-caliber team at the beginning, and that's why we had to rebuild. Rebuilding is an important part of baseball; but once a general manager is finished putting a club together, it is the manager's job to get the most out of his players. That is what occurred in 1979. Because of the years that have passed, Pirates fans may not recall that we had a new third baseman, a new shortstop, a new second baseman, and we also added a few players during the course of the season.

Of course, Willie Stargell was our most valuable player although he may have been older and not playing as much as he did earlier in his career. Having "Pops" on my club was like having 10-karat diamond on my finger, and that is why I told him he would always have a job as long as I had one. I told him he would always be with me because he had the right makeup and made me a better manager. Willie was the type of man who, when he talked about the game, his teammates would listen closely. He was one of the greatest individuals I ever knew and one of the greatest friends I've ever had. He's on my all-time All-Star team. Stargell was a crown jewel, and I have missed him dearly since he passed away on April 9, 2001.

Dave Parker was also very important to the 1979 Pirates. That man could do anything on a baseball field. He could run, he could field, he could throw, he could hit for power and he could hit for average, and that is why I believe he belongs in the Hall of Fame. As a young prospect, a lot of scouts believed Parker was a five-tool player, but on too many occasions young players never fulfill their potential. Thankfully for our ballclub, that was not the case with Dave. He played hard each day, and he played the game with heart. He wanted to win more than most men I've encountered in this great game, and the city of Pittsburgh was lucky to have the chance to watch him play as long as it did.

But Willie Stargell and Dave Parker were just two players on a club that owned perfect balance from the top of the batting order to the bottom of the bullpen. We owned a quality bench and great relievers, and one man who I believe doesn't receive the credit he deserves is Kent Tekulve. That young man pitched in 94 games during the season and then in seven more in the postseason, and never did he ask to be rested. In the playoffs, I used him differently than I had during the regular season. Instead of putting him in the there in the eighth or ninth, he relieved in the sixth and seventh innings because Cincinnati was loaded with the likes of Johnny Bench, Joe Morgan and Ken Griffey. Tekulve used his sinker and slider to hold the lead, and then Don Robinson would go in and use his curveball and 97-mph fastball to slam the door shut. It worked so well that I continued to do it in the Series.

World championship seasons like 1979 do not come along often, but every once in a while the halo sits over your head and magic happens. Baseball fans can research the history of the game and find many, many managers who have never won a World Series, and that is because it is the hardest thing to accomplish in all of sports. No matter what anyone believes, good players make good managers. You have to have good players, and you have to be able to communicate with them. Then, there must be a certain electricity buzzing around the team while that halo is placed over your heads for everything to fall in place. For example, my mother passed away just before the fifth game of the World Series, but I told those guys that she went upstairs to give us some help. I was very upset by my mother's death, but because of the men I had on the club I was able to deal with the sadness and march forward to the greatest achievement a ballplayer can ever hope to realize.

I have always been a believer in one theory: In order to win in baseball, you cannot be afraid to fail. That is why I never tried to avoid losing while managing. You have to try to manage to win throughout the whole season – each day, and every game. You can't be afraid to make the calls you believe are the best in a particular situation, and you can't be afraid to use the abilities of your players in any way you believe will best help the team to get that extra run or the last out. Sure, I got booed, and I heard them loud and clear, but at the same time I was confident that I was making the right decisions.

I remember one time I pinch-hit John Milner, a left-hander, for Steve Nicosia, a right-hander, and because there was a lefty on the mound I got booed a bunch. I even turned toward the crowd to see if my wife was leading the booing. But Milner hit a grand slam, jeers turned into cheers, and the magic continued another day.

I always thought of myself as the strongest man on the club because I could handle any situation on or off the field. All my players were equals. I had to do what I thought was right, and I was always in charge. I was the boss and made every decision during the game, and I still insist that I never made a wrong decision in my life that didn't work. I know most fans believe I always had a smile on my face, but that was not the case. I was tough, but I did things that were necessary. I used to hug Omar Moreno when he made a mistake, but I would scream at Ed Ott. I treated each player differently according to how I read them, and that is how I would communicate with them on an individual basis. Some players needed to be coddled and others needed a kick in the rear every once in a while. In the end, that ballclub played the game the way it was asked, and we were able to bring home a championship to the city of Pittsburgh.

I know I will never be able to repay baseball for what the game has given me. I wear a World Series ring every day only because of a great collection of players who conquered every challenge that faced them. That taught me that, while there are many variables that go into winning, the most important ingredient is for 25 men to play like a team.

I have worked for several major league franchises during my career – playing, coaching, managing and scouting – but all my life I have always loved the Pirates. I have long believed that once you are a "Bucco," you are always a "Bucco," and if you don't believe me, ask any fan or any player who has put on the black and gold. I have love for a lot of things in this world, but baseball and the Pittsburgh Pirates are my life.

The same can be – and should be – said about the author of *Winning Ways*. One of the best weeks I have each year is when I go to the Pirates Fantasy Camp in Bradenton, and I have seen Steve Novotney play the game with a passion that is very rare. He has been a Pirates fan since his father started bringing him to Three Rivers Stadium in the early 1970s, and his love for this franchise today is stronger than ever.

Novotney offers all baseball fans the paths traveled by one of the most storied franchises in the game over the past 50 years. There may be only 14 chapters featuring players and managers who have played key roles with Pittsburgh baseball since Ralph Kiner was an All-Star outfielder, but Novotney has succeeded in capturing the spirit with which every man has played in Pittsburgh. From a pitching pioneer like ElRoy Face to a superstar like Barry Bonds, Pirates fans have witnessed some great baseball, and *Winning Ways* is a testament to the evolution that has taken place before our eyes.

The stories that follow reveal many things Pirates fans do not know about their heroes because Novotney has earned the respect of these former players and many, many more, both as a journalist and as a "baseball man." Steve knows baseball, and although I have admired his enthusiasm and knowledge for many years, he is the first to admit that he remains a student of the game because, as he says, "It is a sport no one has ever mastered completely."

Tales of Past Pittsburgh Pirates
by Steve Novotney

Acknowledgements

The 14 men featured on the pages to follow are atop my list of thank yous because without their shared memories, *Winning Ways* would not exist. These men explain their respective tenures, how particular teammates touched them during their careers, and how victory was orchestrated during those magical seasons in which everything clicked for whatever reasons.

Chuck Tanner quickly accepted the invitation to compose the Foreword for *Winning Ways*, and for that I am absolutely grateful. If he did not believe in the product and the writer, Chuck would not have offered his participation or his recollections about his players and his career. The message delivered in his Foreword is very clear – it takes a team to win baseball's grand championship – and he explains the hows and the whys in Chapter 12.

Charles Jarvis, John D'Abruzzo and John Gworek served as soundboards, editors and researchers for *Winning Ways*, and Beky Davis made this book beautiful beyond the tales this volume contains.

Former Pirates manager Jim Leyland provided much insight on his arrival and years spent in Pittsburgh, and John Perrotto of the *Beaver County Times* contributed much insight, as well. I also would like to thank Jim Trdinich, the director of the Pirates' Media Relations Department, and his staff members, Dave Arrigo, Dan Hart and Patrick O'Connell, for their assistance with locating the photographs that appear.

As a child, I closely followed the 1971 and 1979 world championship clubs, and my baseball hero remains Wilver Dornel Stargell. Thankfully, once accepting the position of editor and general manager of PIRATE

REPORT, I encountered Willie upon his return to the Pirates in 1997 as a special assistant to the general manager. It was Willie's passion for the game of baseball that permitted me to recognize my absolute appreciation for what these men must accomplish to become big-league ballplayers in the first place.

Thank you, Willie, and thank you baseball.

1

Ralph Kiner
Pirates Left Fielder, 1946-53

"The pitchers knew exactly who he was, and not just because he was the only threat we had on our club. He was one of the best in those days, and if a pitcher threw a mistake at him, Ralph Kiner made them pay for it."

– Bob Friend

September 11 had been a day of fond, proud memories for more than a half a century for Hall of Famer Ralph Kiner. After September 11, 2001, though, the veteran New York Mets broadcaster, like all Americans, had something else to remember on that single date. Kiner was in the Big Apple the day suicide terrorists flew two jetliners into the twin towers of the World Trade Center and thousands of innocent deaths were mourned around the world.

In addition to the terrorist attack in New York, the coordinated catastrophe resulted in a third airliner crashing into the Pentagon near Washington, D.C., and a fourth falling into a field in western Pennsylvania as brave passengers tried to regain control of the hijacked plane. Hundreds more were lost in those senseless acts.

Normally when the calendar date changed to September 11, Kiner would recall once again his greatest day as a big-league ballplayer. Seldom, if ever, had baseball seemed so unimportant to Kiner as it did after the carnage of the day now simply known as 9/11.

"If you're an American, you don't care about anything but mourning on September 11th," said Kiner, a three-year Navy veteran of World War II. "It was a tragic day. It was a day that changed everyone's perspective forever. When things like that happen to people you know and to a country you love, you come to appreciate your life more. That's when you really realize what's important and how lucky you've had it. I know I've been a lucky man."

Fifty-four years earlier, the stage was set for Kiner as the talented Boston Braves were visiting Forbes Field. The Pirates and Braves would play four games in three days. Boston was embedded in a heated race for the National League pennant, but the Pirates were staggering toward the season's finish with only 55 wins in 138 games when the Braves came to town.

Kiner, among the league leaders in every offensive category in just his second major-league season, served as the lone highlight on a team destined to finish 30 games below .500 and tied for last place with Philadelphia. On Sept. 11, 1947, however, Pittsburgh proved better than the Braves, and Kiner was the brightest star of all as the Pirates swept a doubleheader. In eight plate appearances, Kiner stroked four home runs and collected 19 total bases, and the Pirates won 4-3, and 10-8.

The next day Kiner swatted two more homers to lead the Pirates to a 4-3 victory. "Most of the time, I've been reminded of that one day while announcing the games," explained Kiner, who began his long broadcasting

Ralph Kiner – 3

career when the expansion Mets entered the majors in 1962. "They would make it the trivia question, or one of my partners would bring it up during the course of the game that day. That was a long time ago, that's for sure, but I still remember it pretty well. You don't have days like that and forget them too easily. To hit six home runs in two days – I don't care who you are – you really have to be seeing the ball well. It didn't get much better than that."

Kiner's major-league career began in 1946 when the left fielder debuted with the Pirates at the age of 23, and as a rookie he showed promise despite registering a .247 batting average while striking out 109 times in 502 at-bats. The outfielder led the National League with 23 home runs, and his 81 plated runs were a team best and fifth overall in the circuit. Although the Pirates compiled only one winning record during his years in a Pittsburgh uniform – in 1948, Pittsburgh was 83-71 – Kiner's reputation grew stronger in the six seasons to follow as the right-handed hitting outfielder became the only major-leaguer ever to win or share the NL lead in homers for seven straight seasons. He also paced the Senior Circuit in total bases in 1947 with 361, plated runs in 1949 with 127, runs scored with 124 in 1951, walks in '49, '51 and '52, and in slugging percentage in 1947, '49 and '51 (.639, .658, .627).

"When he hit them, you knew they were gone as soon as the ball left the bat," said former Pittsburgh pitcher Bob Friend, a teammate of Kiner's from 1951-53. "Ralph used to hit them high and far. They were something to watch, that's for sure.

"The pitchers knew exactly who he was, and not just because he was the only threat we had on our club. He was one of the best in those days, and if a pitcher threw a mistake at him, Ralph Kiner made them pay for it. We weren't that good, not at all, but everyone knew who

Ralph was because of the home runs. That's one thing that hasn't changed – the fans have always loved the home run, and that's what Ralph gave them. The people in Pittsburgh would wait until his last at-bat before leaving our games, and everyone on the team knew that as fact because we saw it ourselves."

Kiner's star status and fan attraction, though, did not stop general manager Branch Rickey from swapping the outfielder to the Cubs in a 10-player transaction on June 4, 1953. The move was reportedly explained by Rickey at the time as necessary for overall team improvement. Kiner, catcher Joe Garagiola, outfielder George "Catfish" Metkovich and southpaw starter Howie Pollet were sent to Chicago for catcher Maurice "Toby" Atwell, reliever Bob Schultz, first baseman Preston Ward, prospect infielder George Freese, and outfielders Bob Addis and Gene Hermanski. The Pirates also received $150,000 in cash from the Chicago franchise.

Only Atwell and Ward appeared as Pirates in 1954, and Freese, a Wheeling, West Virginia, native, appeared in 51 games in 1955. Kiner's career, however, continued past the 301 homers he belted for the Bucs. He was with the Cubs for two seasons before he was traded to Cleveland in November 1954. His playing days ended following the 1957 campaign because of a chronic sciatica problem in his lower back. In his final season, Kiner batted .243 with 18 home runs in 113 games.

Half of a century later, Kiner still despised the trade that sent him away from the steel town he loved and the Pirates fans who loved him. His hero status was legendary, even though the Pirates were 492-739 during his tenure. "It was a great city. Everything was going my way. I owned the town, and you can't have it any better than that," the slugger said. "The one thing that made it somewhat acceptable was the fact that I went to Chicago, which is a great city, too. But I didn't want to

Ralph Kiner – 5

go anywhere. I loved the fans, and they seemed to love me. I couldn't go anywhere without signing autographs and being greeted by so many great people. They loved the home runs."

Long before the era of juiced-up baseballs, Kiner was swatting towering fly balls out of National League parks. "When Mark McGwire came into (the National League) I was asked by a lot of people to compare him to players of the past, but the only person he reminded me of – the way he hit the home runs – was me because he hit high-fly balls that just kept on going like I did. They hit them farther these days because of the physical differences between today's players and the players back then, and also because of the baseball itself. They're using a live ball these days, there's no question about it. There's no question the ball is different."

Unlike the paths followed by the majority of retired players, Kiner remained in the game, first as the general manager of the San Diego Padres for five years when the franchise was a member of the Pacific Coast League, and then as an announcer for the Chicago White Sox. "When I was done playing, I thought I would go into the brokerage business. I played my last year with Al Rosen and he was into it himself. One time when we went to New York City, he arranged for us to go down to Wall Street to meet some people, and I was supposed to go to work as a stock broker out in Beverly Hills.

"But then I got a call from Padres owner (Conrad Arnholt Smith) and he asked me if I would be interested in being the GM of the Padres. I had not given that any thought at all, and it really struck a nerve with me. It sounded great to me. I really liked San Diego, it was near my home in Palm Springs, and as it turned out I ended up being the GM there for five years. After that, I got into broadcasting with the White Sox because Bill

Veeck and Hank Greenberg owned the team. That's how it all started, and now I'm still in it with the Mets. You better believe I consider myself very lucky. The fortunate thing is that I never applied for any of the jobs. They just came at the right time."

His options, as a young war veteran wishing to play a sport for a living in the 1940s, were limited, but the homer-hitting heroes who played before Kiner – the Yankees' Babe Ruth, Lou Gehrig and Joe DiMaggio, Philadelphia's Jimmie Foxx, Mel Ott of the New York Giants and Boston's Ted Williams – divulged an example to follow and the secret to success, Kiner insisted.

"When I was growing up, playing in the major leagues was a dream. At that time, baseball was the only professional sport other than boxing where you could make a living. Football wasn't big at that time, so for an athlete to try to make a living, baseball was it. Baseball was a very important part of my life because I never figured I could make the money I was able to earn as a player. But I did because I did what they wanted me to do, and that was hit for power consistently."

In 10 years, Kiner stroked a home run once every 7.1 plate appearances, second only to Ruth in the game's history, and averaged 37 homers and nearly 102 RBI per year. Williams even selected Kiner as one of the 20 greatest hitters for the "Hitters Hall of Fame" in 1995, and he is still the only player to hit homers in three consecutive All-Star Games (1949-51).

"Being a player, I wouldn't trade it for anything in the world because you learned what it's all about, and I had a great career. I've been in the broadcasting business for longer than 40 years, and it's been an addition to something I've always loved. Everything worked out perfectly."

Though his career was abbreviated due to the back injury, Kiner was finally deemed deserving of induction to baseball's Hall of Fame in 1975. The outfielder compiled a .279 lifetime batting average with 369 home runs, 216 doubles, 1,011 walks, 1,015 RBI and a .398 career on-base percentage. He was an NL All-Star every year from 1948-53; paced the league in 11 different offensive categories, starting with his 23 home runs as a rookie in '46, and ending with 37 long balls and 110 walks in 1952; and was one of 100 former players named to Major League Baseball's All-Century Team in 1999.

"I think anyone who has been in that situation in Cooperstown would tell you that the time you spend at the podium is the best moment in your career. It's overwhelming to be there. 'Maz' broke down when he got up there, and the majority of guys break down or come as close as you can come to breaking down. The only guy I've ever seen go into the Hall of Fame who was sort of blasé about it was Yogi Berra. Maybe it was because he had been in so many World Series, and it wasn't that big of a deal.

"But the majority of us did, and it's because there's no question that it's the biggest sports moment in your life. It's probably the most livable moment of your life with the exception of getting married or having children. That's why I felt the emotion just like 'Maz' did, and why Julian Johnson, who went in with me in '75, had to stop. He held up the ceremony for 15 minutes because it was so emotional for him."

Fifty seasons after his last day in a Pirates uniform, Kiner still possessed several franchise records, including the most home runs hit in a single season (54, 1949), most home runs in consecutive years (111, 1948-49), and most RBI by a right-handed hitter (127, '49). He is second on Pittsburgh's all-time home run list, seventh in RBI and 10th in extra-base hits.

In 1987, the Pirates franchise honored Kiner by officially retiring his No. 4, making him the seventh player in the team's history to receive such an honor. In April 2003, a sculpture of his hands gripping a Louisville Slugger was erected beyond the left field wall of PNC Park.

"When they retired my number, it was a great experience and I enjoyed that weekend very much," he remembered. "When they presented the sculpture of my hands to me on the field, it was a touching time. It was 50 years after I left Pittsburgh, so it meant a lot to me to be remembered by the franchise after that long. I thought that was pretty impressive."

Kiner appeared in 1,472 games as a player, but has broadcast thousands more while witnessing the evolution of a game that has changed its stripes more often than any other professional sport. He played against Brooklyn's Jackie Robinson after the Georgia native dared to break baseball's color barrier in 1947; the adoption of free agency in 1976 has been followed by seven work stoppages; expansion swelled the Senior and Junior circuits from 16 to 30 teams in the five decades that followed his rookie campaign; two waves of ballpark construction have razed many of the ballparks in which Kiner played; and player salaries have dramatically increased since the outfielder became the first National League player to request a $100,000 contract in 1952.

"People ask me if I regret not playing in the big-money era. I say no. I wouldn't change a thing because I stayed in baseball all my life," he said. "It's been a great career, and I've lived through a lot. There's been a lot of change since I first stepped out onto Forbes Field, and I've enjoyed watching it all."

Along the way, Kiner has become a decorated broadcaster while working for two television stations, three cable networks and six different radio stations. He has collected three Emmy awards, the William Slocum

Award from the New York Baseball Writers of America in 1990, and in 2002, Shea Stadium's television broadcasting booth was named in the slugger's honor. Kiner, in fact, was set in late 2003 to publish a sequel to his first book, "Kiner's Korner: At Bat and On The Air – My 40 Years in Baseball."

"The new book is on my whole career, on the greatest teams and everything I've watched occur in baseball history – the arrival of black players, the unions, the strikes, and everything else. I lived through the whole thing," he explained. "We have some letters in there, and one is from me to Branch Rickey. In it, I said I hated to be traded, but basically I had no emotion about that because it's about being in baseball. Everything worked out for me. I can say that now."

The future of baseball, Kiner believes, is bright although the industry was infected with economic strife in mid and small markets during the 1990s and early 2000s. The Hall of Famer suggests even more alterations, but believes America's pastime will continue to grow instead of contract. "I think baseball is going on and on and on. It's a great game, and I think it's going to get bigger and better.

"The game's changed, there's no doubt about that – but you can't beat the brand of baseball that's being played these days. I'm sure more changes will take place, and I think they'll increase the number of games, maybe go to the best five out of eight in the World Series like Pittsburgh played in 1903.

"Life is changing and baseball is changing with it," said Kiner. "It has to, but how can you beat the excitement it's given the fans? It's been fantastic."

The Pirates lost 112 games in 1952, Kiner's last full season in Pittsburgh, and 206 more defeats followed in 1953-54. But in 1960, the "Battlin' Buccos" were born and crowned world champions after defeating the Yankees in seven games. New York's powerful lineup, stoked with the likes of Mickey Mantle and Roger Maris, outscored Pittsburgh 55-27 in the Fall Classic. A single Bill Mazeroski swing, however, handed the Pirates franchise its first title since 1925.

Kiner feels no connection to the '60 ballclub, and insists he owns no reason to claim he helped transform a failing franchise into a champion organization. "I wasn't around when that happened. I was long gone and out of baseball by the time the Pirates became known as the 'Battlin' Buccos'. They earned that nickname for coming back late in so many games in 1960. If it's true that I helped bring that kind of 'never-say-die' attitude to Pittsburgh, fine, but I believe the credit for that lies with the men who followed me.

"The records don't lie – we didn't win much during those days. But yeah, we battled and we never quit. Quitters don't deserve to be ballplayers."

2

Robert Friend, Sr.

Pirates Starting Pitcher, 1951-65

"He was our workhorse, and we could count on him. You don't see pitchers like Bob Friend anymore. That takes guts and good health ... and more guts."

— Bill Mazeroski

Bob Friend proved to be a building block when former General Manager Branch Rickey added the right-handed pitcher to the Pirates' roster in 1951. Friend's debut coincided with Rickey's youth movement that featured premature promotions, roster turnover and the trade of Ralph Kiner to the Cubs in 1953. Friend, a rookie at 20 years old, was soiled, seasoned and durable by the time the Pittsburgh club was ready to snap a nine-season losing streak that included three straight years (1952-54) in which 100 losses and last-place finishes were realized.

Only ElRoy Face remained with Friend on Pittsburgh's pitching staff when Danny Murtaugh was named Pittsburgh's 27th manager in 1957, and that starter-closer combination played a vital role in the ballclub's winning seasons in 1958-59. "A lot of us were very young. I was only 20 years old when I joined the team. I probably wasn't ready," Friend explained. "I wasn't a major league pitcher. I was on the major league roster, but I was not a major league pitcher. It took me about four years to really get zoned in getting to pitch on a regular basis.

"We had a bad ballclub, and I wasn't a very good pitcher, and any of the young players who came up – most of them in that day – were not ready. But (Branch) Rickey kind of pushed it. He'd say, 'If we're going to lose, let's lose with the young players.' I stayed the course, and in 1955 I had that great year to win the ERA championship (2.83) and win 14 games. That's where I really gained my confidence. That was the start of my career."

Former managers Billy Meyer and Fred Haney both used Friend as a starter and reliever, but in 1956, new skipper Bobby Bragan quickly regulated the hurler to starter status. That season, Friend tossed 314 1/3 innings and recorded a 17-17 mound mark, 19 complete games and a 3.46 earned-run average. "Bragan worked me the way I liked to be worked – two days rest, three days rest. I was 11-6 at the All-Star Break and I pitched in the All-Star Game and got credit for the win. I thought I was going to win 25-30 games.

"Well, that didn't happen. I might have gotten a little tired at the end of the year, but that was the way it was," he said. "But I liked pitching on two or three days rest. I was able to do it because, I guess, I had a rubber arm."

No matter which manager was completing the lineup card, Friend never turned down the baseball in his 16 big-league seasons. In 15 years in Pittsburgh, he compiled more frames hurled (3,481) and more strikeouts (1,682) recorded than any other man to pitch in a Pirates uniform. Friend is third in total appearances with 602, third in shutouts with 35, fourth in victories with 191, seventh in complete games with 161, and he also sits

Robert Friend – 15

atop franchise records with the most walks (869) and losses (218).

"Bob Friend became the kind of pitcher every guy wants to become," said Hall of Famer Bill Mazeroski, a teammate of Friend's from 1956-65. "He established himself as a guy you could count on, and he gave his teammates the feeling that they would have a good chance to win with him on the mound.

"He was our workhorse, and we could count on him. You don't see pitchers like Bob Friend anymore. That takes guts and good health ... and more guts."

Friend, who admitted he did not own a special offseason training regimen to ensure arm strength, simply believes he was lucky. "I never missed a start, that is true, but I didn't throw the ball all winter, either. I never threw it. I never picked up a ball except during the season; from spring training, I threw every day. Now, I have pitched with pain, but it wasn't enough to get me worried about it. I did have an elbow problem at the end of 1954, but it was right at the end of the season and I had all winter to rest. It never came back.

"Back in those days, they didn't have the technology or the know-how they have now," said Friend, an insurance broker for 25 years following his retirement from baseball. "There were a lot of guys who would hurt their arms, and you'd never hear from them again. They'd have to go get real jobs and forget about baseball. These days, they disappear for a year, and then they came back better than they were before. When I remember that, I know how lucky I was never to suffer an arm injury so I could be part of something great."

Although the righty paced the National League in 1958 with 22 victories, it is the 1960 season Friend recalls most often. "We won it all, and because of the way we beat the Yankees," Friend said. "I know most people remember 'Maz's' home run the most, but that season and that Series was really special to everyone on that team.

"That's the year they started calling us the 'Battlin' Buccos' because we won so many games in the late innings. We got behind, but we had guys come off the bench and get the hits. It was a tremendous year. It was the best baseball year of my life."

Face, in fact, believes it was Friend's contributions in 1960 that permitted the Pirates to win the National League title. "I'm not sure if he had a bad game that season," the reliever said. "As the guy who went in after the starters were done, I remember knowing that there was a good chance I wasn't going to get to pitch on the days Bob started. He pitched the same game no matter who he was facing, and that's what pitchers used to be made of."

Friend won 18 games and lost 12 in 37 starts during the championship season, registering 16 complete games and four shutouts while the Pirates compiled a 95-59 record. He whiffed 183 and walked only 45. "If you look at the statistics, I think I pitched better than ever in 1960. I was five-to-one on strikeouts to base-on-balls. I had 37 starts and probably had 35 quality starts that season," Friend said. "Now, I wasn't that good in the Series (0-3, 13.50 ERA), but that's what was special about that team – other guys stepped it up when they had to, and it all worked out.

"We were proud to be called the 'Battlin' Buccos' because that's how we played the game. We were never out of the game until someone told us it was over, and that's why the Series ended perfectly. Just when the Yankees thought they had us down for the last time, 'Maz' goes up there and the rest is history."

Friend was 29 years old and a 10-year veteran when the city of Pittsburgh celebrated the Pirates' first World Series victory in 34 years, and after five more seasons

(70-79. 3.70 ERA), the hurler was met with a career crossroads that ultimately led to pitching for the Big Apple's Yankees and Mets in 1966 – and to the end of his baseball career. "I turned down what (the Pirates) wanted me to do, and that was to quit being a starter and start being a reliever," he said. "And I wasn't showing any enthusiasm about going into the bullpen after starting. It was a pride thing; I'll admit that now.

"I was 35 years old at the time and believed I could still start and wanted to prove it. That's why I left for New York City – and it was a great experience. I thought we were going to have a good ballclub, but in reality you could see the guys were slipping a little bit. (Mickey) Mantle was hurt and, of course, (Whitey) Ford had the bad arm. The guys were getting old and we just didn't have the club. It was great to be there. I got to meet some of my heroes, like Joe DiMaggio, who was a coach in spring training.

"Then the Mets bought my contract in June, and I went over there and met some nice guys. I got off to a nice start, winning my first couple of starts, but the writing was on the wall. After all the starts and all the innings, I was done. Who knows? Maybe I'd have been able to pitch a few more years if I had gone to the pen, but there's no use of thinking about that now."

The right-hander owned the nickname, "Warrior," in his playing days, but not only because of his approach on the hill. Friend served as an important cornerstone when on-field personnel initiated organization in 1950s.

"Money wasn't the problem back then. The things that we were concerned with involved traveling on off-days, traveling after night games and things like that. (Hall of Fame pitchers) Robin Roberts and Jim Bunning came to me and thought there should be change with the players. They wanted to get a stronger man as legal representation who was concerned with the players' welfare.

At that time the pension-retirement program wasn't much at all because the only money we were getting was from a percentage of the All-Star Game.

"We felt we needed someone in the job full time so we formed a committee to look at it. We even had (former President) Dick Nixon and his law firm look at it when he was out of office after losing the governorship in California, and he turned it down," Friend recalled. "(Kentucky Senator) Bunning knew Author Goldberg, who was the labor man under President Lyndon Johnson in 1965, and he suggested Marvin Miller."

Miller was hired by the players in 1966 and immediately approached commissioners Bowie Kuhn and Spike Eckert with the experience he gathered while working with the steelworkers' union. During his 18 years as the players' union chief, Miller negotiated several victories, including an end to the reserve clause so to permit free movement from team to team through free agency; arbitration in labor disputes; the right for veteran players to refuse trades; an improved pension plan funded largely through percentages of television revenue; and, first, recognition of the players' union, the right to bargain collectively, and the use of agents to negotiate individual contracts.

These Miller triumphs, however, led to a pair of work stoppages, first in 1972 and again in 1981, and effectively created a system in which the game's elite players were free to move from team to team, eliminating a sense of loyalty with the fans and defining baseball as a business more than ever before.

"Back then, there were a lot of players and fans who thought creating a union would bring down baseball, but that wasn't our intent at all. We just wanted to be treated fair, that's all," Friend said. "We took a lot of heat on it, but we didn't know it was going to go the way it has since. I think we opened the door for a lot of good things for the players, but we know now that not everything has worked out positively, and that some of the things have been taken too far.

"Without the union and what Marvin Miller did, the pension program would not be one of the best there is today, and back in the 1940s there were a lot of players who never could take advantage of opportunities with other teams because free agency wasn't around. Plus, there were some things that the owners were allowed to do that just were not fair."

Friend, in fact, found himself situated in a payroll decision made by general manager Brown. After posting an 8-19 record in 1959 with a 4.03 earned-run average – his highest ERA since posting a 5.07 in '54 – the hurler received notice his compensation would be decreased by 25 percent. "I got off to a bad start and Murtaugh kept pitching me. I wasn't throwing the ball well early and often enough throughout the season," he explained. "I don't know why, but those things happen and Murtaugh stayed with me. I had some good games in the end, but I still got (a pay) cut.

"Hey, I accepted it because I knew I had a bad year, and because it was part of the game back then, fair or unfair. And I came back, got the raise after the 1960 season, and a raise above that the next year because I always believed it was about more than how much money you made playing the game. Sure, I wanted to make enough to survive, but I wanted to earn it. Now, almost 40 years after I quit playing, I can honestly say that I earned every dollar I made playing baseball.

"When we were in our fight, we believed we were doing something good for the game of baseball. There were a lot of things that needed to be changed, and getting all the players in on it was very, very important. But where it's gone since?"

Only a few rules between the lines have been altered

since Friend's retirement in 1966, including the lowering of the pitcher's mound in 1969, and the implementation of the designated hitter in the American League in 1973. Off the field, though, Friend has witnessed much reconfiguration with the game's economics and the methods of management when handling high-priced personnel.

"The baseball today is great, but it's different. The fact that they have so much money involved with some of these pitchers' arms – they don't want to take a chance with injuries by making them work as much as we were asked to when I played," Friend said. "They know much more about the human body these days, so I'm sure that plays a big part in the decisions that general managers and managers make, but if you look at the statistics from the 1950s and 1960s, and compare them to the numbers of the 1990s and after the turn of the century, you'll see a big difference. These days, there aren't too many guys throwing nine innings. They've got specialists now – middle relievers, set-up guys and closers, and apparently it's working."

Over his 16 seasons, Friend averaged 226 frames pitched per season and surpassed the 250-inning plateau six times. In 2003, only eight major-league pitchers tossed more than 225 innings with Toronto's Cy Young Award winner Roy Halladay pacing all with 266. "It's unheard of these days for a starter to pitch as much as Bob Friend used to," former Pirates pitcher Nellie Briles said. "First, there used to only be four-man rotations when there were less teams, but that didn't mean the guys pitched less. They pitched more.

"It was the same all around baseball in those days – you pitched until you couldn't anymore. Some guys got hurt and would either go to the bullpen or be gone, but Bob kept coming back because he was blessed with one of those arms," Briles continued. "These days, though, it seems as if every pitcher suffers injuries and they miss time while getting surgery and doing the rehabilitation. If you compare the era when Bob pitched and today, the differences are huge when talking about the role of a starting pitcher, and the money involved."

The same can be said for all professional sports, however, what with the advancements made in medical technology, the evolution of the American economy, and the inflation of pay for the pro athlete. Friend has been touched by the effects – his son, Robert Jr., has performed as a professional golfer on the Nationwide and PGA tours since 1990, and earned more than $1.3 million during his first 14 years.

"But there's a big difference between baseball and golf these days, isn't there?" Friend insisted. "In baseball, there are guaranteed contracts so a player makes his money if he's hurt, and even if he performs well below what's expected. My son, on the other hand, makes money if he earns it, and he knew that going in. You have to build up a tolerance for that, and he has built up a good tolerance. Sure, the golfers earn more today than they did when I was his age, and he's done pretty well with it – but he's got to be healthy and playing well to make a buck. It's that simple, and if baseball could get to that point, I think it'd be better off."

He follows the game, and in particular the Pittsburgh Pirates, and Friend, a regular linkster at Oakmont Country Club, believes steps have been taken toward correcting a plethora of problems created since his playing days and the inception of the players' union. "But the smaller-market clubs still need some help. Whether or not the bigger-market clubs want to do that, I don't know. The smaller-market teams can win, and that's been proven, but they also haven't been able to keep their best players because of the way the system has been set up. If there aren't some changes made soon,

I think we'll see some teams disappear, and that would be a shame."

3

ElRoy Face

Pirates Closer, 1953-68

"ElRoy helped usher in change in baseball. He was one of the first closers and is still considered one of the best closers in the history of the game. All I did was follow his lead."

— Kent Tekulve

Most of ElRoy Face's weekends have been spent with his wife, Roberta, on a small chunk of land in Elderton, Pennsylvania, a tiny town located between Indiana and Kittanning, since his retirement in 1989 – not from baseball, but from a carpenters' foreman position at Mayview State Hospital in suburban Pittsburgh.

Following a playing career that spanned 16 seasons and 1,375 innings pitched with Pittsburgh, Detroit and Montreal, Face went to work, blue-collar style. "I had to – most of us had to get jobs when we were done playing. We didn't make the money they make today. These days, a baseball player can sign a contract for a couple of seasons and be set for life.

"Playing baseball paid the bills for almost 20 years, and that was great. I consider myself very lucky. But when I was done pitching, the bills kept coming so I had to do something. Some guys went into business selling insurance or stocks or whatever. I was a junior carpenter when I was playing so that's the direction I went in," Face explained. "You know, Bill Mazeroski is the only guy I know who's never had a real job."

It was Mazeroski, of course, who hit *the homer* clearly heard throughout the game. All Face did was silently set new standards for relievers while pioneering a new position in baseball: the closer. Not until his final season in 1969 did Major League Baseball officially recognize a new statistic, "Saves," and after records and box scores were examined, Face owned 193 – 188 of which he compiled while a Pittsburgh Pirate.

Yet, while Mazeroski and his Hall of Fame name survived on signature circuits and golf courses, Face fell back on what he knew best. First, the right-hander was into the bar business, and then he built and installed aluminum awnings and replacement windows. Before taking the position at Mayview in 1979, he worked as a contractor building houses. "And since, I've spent a lot of time in my motor home camping, fishing and playing quite a bit of golf," he said. "I can't complain because it's all worked out OK.

"Baseball hasn't been a big part of my life since I quit playing. The only involvement I've had with the Pirates has been with the Alumni Association and playing a lot of charity golf, and that's always a lot of fun because I get to see many of the guys I played with, and talking with the fans has always been fun for me."

Face, though, often recalls his closer career, especially when watching the Pirates at PNC Park or when the ballclub's games are televised. "When you spend 20-some years in the game, it never leaves you. It's a part of you. I watch the hitters and the pitchers, and I'm always thinking about how I would have thrown to this guy and

ElRoy Face – 27

that guy – up and in, down and out, with this pitch or that one. I'm still pretty good at it."

The three-time All-Star's name appears throughout the Pirates' record books, and Face also owns National League records still standing although specialization has increased in baseball during the four decades that have followed his final day in uniform. His 188 saves are the most registered by a Pittsburgh reliever, and his 802 appearances ties him with Walter Johnson for the most games pitched for one team. He was 104-95 overall with a 3.48 career earned-run average.

In 1959, Face strutted a string of 17 consecutive wins and captured 18 victories in relief, both of which remain big-league bests. He also appeared on national television as a guest on "The Ed Sullivan Show."

But Face feels he was at his best in 1960 when, in an NL-leading 68 appearances he posted 24 saves, a 10-8 record and a 2.90 ERA in 114 2/3 innings. "That's when I really started getting my forkball to do what I wanted it to. It had a lot of movement and I got a lot of ground balls when we needed them," he explained. "In 1959, I'd go in with a lead but they would tie it up on me. Then we'd score again, and I'd get the win instead of a save.

"In '60, though, they didn't tie it up very often. I usually got them out and held them down where they were when I got in there."

Until Game Seven of the World Series, that is. Face registered three saves while surrendering just three runs in nine frames in games one, four and five, but soon after entering the Series' final game in the sixth, the Yankees scored three times and starter Bob Friend replaced Face after the eighth. "But the guys on that team never quit, and eventually we won the game. We were the toast of the town. None of us could go anywhere without being surrounded by fans looking for autographs and handshakes."

Face's face is still recognized on occasion in Pittsburgh, but seldom has he been invited to memorabilia shows as a featured Pirate of the past. "Well, I don't have an agent so that's why I don't get as many offers as some of the other guys do, but I've been at a few of them and enjoyed them.

"I'm recognized most places. One time, a couple of years ago, a man recognized me on a plane and he told me he remembered me from 1960 when I was the only guy on our team who would sign an autograph for his kid," Face continued. "That's why I always signed for anyone who asked. I don't remember ever turning down a request, and when that man told me that story, it made it all worth it.

"I also belong to the VFW club in McKeesport, and all the folks there remember my playing days and those teams. It was funny though, when this one day a sales associate from the Pirates called there at the VFW to try and set up a group of members to go to a game. After the commander told this young man that I was a member, he said, 'Who's that?' I guess not everyone remembers."

The Pittsburgh franchise acknowledged Face and his career during the 1994 campaign, the season during which the city hosted baseball's All-Star Game, by naming him the Pirates' best-ever reliever over the likes of Kent Tekulve and Dave Giusti. "What a great honor that was," Face said. "But then, when Major League Baseball named their top 100 in 1999, they named Tekulve the best reliever of the 20th Century, so who knows what to think."

Tekulve, who joined Face in 1979 by recording three World Series saves as Pittsburgh beat Baltimore in seven games, recorded 158 saves in 11 seasons with Pittsburgh. Tekulve is proud of his accomplishments and what honors followed, but is quick to acknowledge Face as a trailblazer for relief pitchers. "ElRoy helped usher in change in baseball. He was one of the first closers and is still considered one of the best closers in the history of the game," the former submarine-style hurler said. "All I did was follow his lead.

"He was one of the first relief pitchers who was asked to adapt, and to be in there at the end when the game was on the line, so I have a lot of respect for that because I know how he felt out there."

Face, however, feels the deeds he, Tekulve and other past Pirates and players logged will fade in the fans' memory more and more despite being repeatedly chronicled in baseball encyclopedias each year. "I'm sure, as more seasons pass, what the players and teams did when I played will be forgotten. That's just how it works. When you look at the closers they have these days, who can blame the fans for forgetting?

"The history of the game isn't as appreciated like it used to be, and the strikes and the all the money problems are probably why. The fans don't want to hear about all that, they just want to watch good baseball."

That, however, does not mean Face believes he and those he competed with and against in the '50s and '60s can now be considered inferior to the ballplayers of the 21st century. His opinion, in fact, is quite the opposite.

"A lot of the younger group don't remember the older group. I think in the 1950s and '60s we had the best group of players in major league history – the (Stan) Musials, the (Mickey) Mantles, the (Willie) Mays, the (Hank) Aarons, and you can go on down the list. Every club had three or four very good hitters and two or three very good pitchers. Now, they are spread so thin because of the expansion that's taken place, they have Double-A players playing up at the major leagues."

Along with Musial, Mays and Aaron, Face joined NL All-Stars Ernie Banks, Orlando Cepeda, Don Drysdale, Eddie Matthews, Frank Robinson, and Warren Spahn to battle against the America League's Mantle, Roger Maris, Yogi Berra, Harmon Killibrew, Brooks Robinson and Ted Williams from 1959-61

"Aaron, Musial, Matthews, Robinson – every club had guys like that. There were more quality ballplayers in the '50s and '60s. Before us, there were the Honus Wagners, the Babe Ruths and the Ty Cobbs, but they were few and far between. In the '50s and '60s, every club had two or three guys who could really hurt you at the plate, or shut you down on the mound.

"That's why there's a part of me that wishes baseball would go back to the 16 teams, eight in each league. I think it's gotten too big and that there's not enough good players," Face insisted. "But that's because I remember those guys like it was yesterday, and that's not the case with most fans these days. They don't know any better, so it's not a problem."

It is the same game, Face concedes, no matter who dons the uniforms or own the teams. "Baseball hasn't changed – they still have to throw the ball, and hit it and run. But the players and the owners have changed, that's for sure. They go about things a lot different than we did, with all the training, all the money that's out there now, and how they treat the fans.

"The money has definitely had something to do with it, for sure. The players have become more independent because of all the money, and they forget that it's the fans who are the ones paying those huge salaries in the first place."

The reliever's rookie season was 1953, and at 25 years old, Face joined the Pirates as a Rule 5 selection following farm system stints with Philadelphia and Brooklyn and earned the baseball's minimum salary – $6,000. After registering his first 75 saves in the seven seasons that followed, Face was one of Pittsburgh's most expensive players, but that did not mean he was exempt from salary slashes when his performance proved under par.

"My highest salary was in '61 (6-12, 17 saves, 3.82 ERA) and it was $42,500, and I think at that time I was one of the highest paid players on the ballclub," Face recalled. "Then I had an off year in '61 so they cut me $5,000. Then in '62, I won the (National League) Fireman Award, and the owners wanted to give me a $2,500 raise. But I held out and got the $5,000 back.

"I'm not mad or bitter about the money the players make these days. I just wish I were 30-40 years younger. Back when I played, $42,500 wasn't a bad salary at all compared to the others. I think (shortstop Dick) Groat was making $25,000. I may have been one of the highest players on the club before (Roberto) Clemente started coming up and getting his raises. He started getting better and better, and because of that I think he became the first $100,000 player on the club."

No matter what records he owns, the path paved for future pitchers, the cash banked playing baseball, or who remembers what, Face rests easy these days on his Western Pennsylvanian campground free of regret. "If people want to believe I was some kind of pioneer when I played just because of when I was usually put in to pitch, that's fine with me. I didn't think much about it back then, but apparently there were a lot of people who did."

That is why, when the time arrived for Pirates GM Joe L. Brown, manager Larry Shepard and owner Dan Galbreath to bid Face farewell, they orchestrated an unusual move to ensure the reliever would tie Johnson's record for most appearances for one team. On Aug. 31, 1968, starter Steve Blass recorded the first out against Atlanta. Shepard then called a time out, shifted Blass to left field, and summoned Face from Pittsburgh's pen. Face promptly got second baseman Felix Millan to ground out, was removed from the game, and then was sold to Detroit.

"I appreciated that gesture very much, I really did because it meant a lot to me to be a Pirate," Face said. "The deal that sent me to the Tigers didn't make much of a difference to me because I knew I was coming to the end of my career. I pitched one more year in Montreal and then it was time to call it quits when the Expos released me in August of '69. That's when it was time to go back to work.

"It was a long career, and it was a good career, and I had a chance to be part of something very special. Take 1960 – there were a lot of people who didn't think we deserved to be on the same field with the Yankees, let alone in the World Series against them. That franchise has always had an edge as far as money, but when it came right down to it, it was all about heart and soul. We battled and never quit, and in the end we proved that they put their pants on the same way we did. We just did it a little better."

4

Bill Virdon

Pirates Center Fielder, 1956-65, 68

#18

"There have been two men I've come to trust more than anyone, and those two men are my father and Bill Virdon. My father taught me about life, and the 'Quail' taught me baseball."

— Lloyd McClendon

Fifty-two years had passed by Bill Virdon's eyes when he made the decision to walk away from the everyday rigors of baseball. His September 2002 exit was on his terms, an option not often presented to those who choose to live a life according to the game.

Virdon had served a two-season tenure as bench coach for Pittsburgh manager Lloyd McClendon when, at 72 years old, the former center fielder decided to retire following an 11-year career as a player, 13 campaigns as a big-league manager, and five different homecomings to western Pennsylvania. But Virdon, known for few words and a hardnosed work ethic, insisted his half-century of hands-on experience had not revealed solutions to the game's most elusive mysteries.

"You don't know why someone decided to invent the

game," he said. "And there's no way to learn all of it. You can learn the situations, but the game, and the other team, still manages to beat you. You can know how to win, but that doesn't mean it's going to happen, and it never happens all of the time. The only thing I know for a fact after all the years in the game is that every year something happened that I'd never seen before.

"And you don't know when you make the decision to make the game a big part of your life because it's not your decision to make," continued Virdon, who was nicknamed "Quail" by broadcaster Bob Prince in 1959. "All I knew was that I wanted to stay in it after playing, and I think I've been real lucky to be able to do what I've done to make a living. I've always tried to stay true to the game. I know there aren't a lot of people who think about that, but at least I always knew where I stood."

Before his final return to Pittsburgh, Virdon accepted a position in 1997 as bench coach for the Houston Astros, the same club he skippered for eight seasons in the late 1970s and early '80s. Larry Dierker, a 14-year player with Houston and St. Louis, jumped from the Astros' broadcast booth to the dugout, and Virdon was hired for guidance. Houston, in fact, battled until late September before taking the National League Central Division title from a surprising Pittsburgh club, a team that won 79 games despite a big-league low payroll of $9.5 million.

"Larry was as sharp as about anybody I ever helped," Virdon explained. "He did play for me so I knew he was intelligent and that he knew the game, but when you haven't managed, things happen on occasion that slip up on you like signs and things like that.

"He caught on, and it was a fun year because we ended up winning the division over the Pirates when their ballclub was playing much better than anyone expected. After that one year, I retired again."

Bill Virdon – 37

Former Pittsburgh general manager Cam Bonifay hoped for much the same in 2001, the ballclub's first campaign in PNC Park. Bonifay replaced Gene Lamont with McClendon, and the '01 season was expected to present the return of winning baseball to a city that had suffered through losing seasons each year since 1992. Virdon, at 70, opted once again to shed retirement to assist a former player.

"I came back one more time because of the game, and because of Lloyd," he explained. "When 'Mac' played for me, we talked a lot so I knew how much he loved the game, that he was as sharp as they come.

"I've not seen too many players who worked as hard as he did before he retired, and that was because he was trying to make himself as valuable to our club as he could. The man just wanted to win, and he brought that to managing when he was hired."

Winning, however, did not prove to be the theme of the 2001 Pirates. Instead, Pittsburgh was wrecked by 19 disabled list assignments and more than 1,100 days lost to injuries, resulting in its first 100-loss season since 1985. "That year wasn't 'Mac's' fault, not at all," Virdon said. "No team – not even the Yankees – could have survived that kind of season. You can't say that it wasn't fair because injuries are a part of baseball, but you also have to be realistic."

Virdon surprised McClendon by accepting the manager's offer to return in 2002. "What manager wouldn't want a man like Bill Virdon to lean on?" McClendon said. "The (2001) season was a tough one to take for all of us because we knew what we could have been. I'm sure not wanting to go out after losing 100 was part of his motivation. Whatever the reasons were, I was thankful.

"There have been two men I've come to trust more than anyone, and those two men are my father and Bill Virdon. My father taught me about life, and the 'Quail' taught me baseball."

Virdon first arrived to Pittsburgh following a pair of transactions that sent the Michigan native from the Yankees' organization to the Cardinals in 1954, and then from St. Louis to the Pirates' outfield in May 1956. The outfielder had been named the NL's Rookie of the Year in 1955 after Virdon batted .281 with 17 home runs and 68 plated runs, but Cards GM Frank Lane swapped him for pitcher Dick Littlefield and outfielder Bobby Del Greco. At the time of the trade, Virdon was struggling with a .211 batting average. "I'm sure that had something to do with the Cardinals trading me," he said. "There have been a lot of players who had good rookie seasons but weren't worth much after that for whatever reasons."

Virdon proved Lane wrong once joining a young Pirates team that, despite its 66-88 record, featured a strong-armed right fielder named Clemente, a slick second sacker known as "Maz," and a pitching rotation with Bob Friend and Vern Law. "There's no doubt that a team was being put together in the late '50s, and because I bounced back that season (batting .334 in 133 games), I was able to be a part of what was to happen in Pittsburgh."

In 1958, Pittsburgh snapped a nine-year skid with an 84-70 mark, and above-.500 play continued the next season. But in 1960, the "Battlin' Bucs" were born. Not since 1925 had the Pirates claimed baseball's title, but a 26-22 record in one-run games was an important ingredient to a 95-59 campaign, a National League pennant, and an improbable World Series championship over the Yankees.

"There were a lot of days that year when things just worked out. It's tough to explain," Virdon recalled. "When we needed a hit in a critical situation, it seems as if we always got one – and not just in the Series, but all year. That's where the nickname came from, and why our fans usually didn't leave our games without us getting the final at-bats."

It was, of course, the Pirates' final turn at the plate when infielder Bill Mazeroski stroked Ralph Terry's 1-0 slider over the left field wall of Forbes Field for the title, but the second baseman's heroics would not have been necessary if not for the dramatics that played out from the game's first pitch. Pittsburgh claimed an early 4-0 advantage after the first two frames, but the Yankees tallied seven runs in the fifth, sixth and eight innings to secure a 7-4 lead.

A freak hop on Virdon's tailor-made double-play grounder in the bottom of the eighth struck New York shortstop Tony Kubek in the throat, and instead of two outs and no one on base, the Pirates owned a pair on the base paths with no outs and Dick Groat, Clemente and pinch hitter Hal Smith coming to the plate. Five runs followed, and Pittsburgh held a 9-7 lead.

"But that lead didn't hold up either, did it?" Virdon said about the Yankees' two-run rally in the ninth. "We weren't too happy with ourselves, but even though they tied it up we knew we had the chance. We'd done it all season long so this ninth inning wasn't going to be any different than before. We knew we had a chance."

With infielder Dick Stuart on deck and Virdon set to bat third in the frame, one Mazeroski swing settled the score at 10-9. "As soon as you heard the sound, you knew it had a chance, but I think we all paused before we celebrated because we wanted to make sure what we just saw actually did happen," said Virdon, who hit .241 with seven hits and five RBI in the Series. "It definitely is the best moment I can remember, as a player or a manager, and I consider the whole 1960 year the best baseball season I've experienced. That was a special year.

"The best part was the collection of players we had on that club. There was no one leader because it seemed we were all about the same age and able to take care of what we needed to do to get our jobs done. The same feeling was there the year before, the year after, and other years, but in 1960 I think everything you always hope comes together actually did."

Virdon added to his hardware in 1962 when he was named one of three Gold Glovers in the NL after contributing 10 outfield assists and a .976 fielding percentage in 156 games. "That was the first year they didn't give them out according to position – they just gave the awards to who (the writers) thought were the best three outfielders," he said. "You think I would have ever won a Gold Glove as a center fielder with a player like Willie Mays playing the same position in the same league?

"I felt I had better defensive years than I did in 1962, but there are things that are just out of your control. I know that I didn't play the outfield trying to avoid making errors. I played to make plays."

Mazeroski, inducted into the Hall of Fame in 2001 after collecting eight Gold Gloves in 17 seasons, produced similar offensive numbers to Virdon during his days in Pittsburgh black and gold. "Back in those days, there wasn't so much concentration on the offense – playing good defense was just as important, and Bill concentrated on that part of his game just like I did," Mazeroski said. "In those days, you had to figure out how you could best help the team. For some guys, they made the most difference with the bat. Some, like me and Bill, helped the most with our gloves, and it was respected back then more than it is these days."

Virdon, though, decided to retire as a player following the 1965 campaign despite batting .279 — 12 percentage points higher than his career average — in 135 games and 481 at-bats. "I caught myself wondering out there," he admitted. "My concentration wasn't there. If I had not wanted to stay in the game, I would have continued

Bill Virdon – 41

to play. I hadn't been able to concentrate like I always had, and I caught myself wondering about this or that and the other when the game was going on and I just didn't want to play like that.

"It was a personal decision, and when it came down to it I knew I had to be honest with the game. After that, I decided I wanted to stay in the game but I knew I would have to do it as a coach or a manager. When I got the chance to see if I could manage, I took it."

That choice, however, took Virdon away from the Pittsburgh organization for two seasons while he guided Class AA Williamsport and Triple-A Jacksonville of the Mets organization before his 1968 return to Pittsburgh as a member of Danny Murtaugh's staff. Three years later, the "Quail" collected his second world title as the Pirates' batting and outfield instructor when the Bucs beat he Orioles in seven games.

"There's a big difference between the two (titles) because it's more fun to compete and play because you feel like you have an active part when you're playing in the games. There's nothing wrong with coaching or managing, but I always felt like I enjoyed playing more than I did coaching or managing.

"When you're playing, you worry about yourself and trust your teammates to do the same, but when you're managing you have to worry about all 25 guys. That's the biggest difference."

When Pittsburgh opened its title defense in April 1972, it was Virdon, not Murtaugh, who served as the Pirates' skipper. Murtaugh, who died of heart problems in December 1976, resigned his position only to hand it over. "Danny set me up for that job because he took me under his wing as soon as I came back and joined his staff. He wasn't in perfect health, but he wanted a break from it all more than anything. After winning again, I think he thought he had done enough."

42 – Bill Virdon

The 1972 season progressed as if Pittsburgh was destined to defend its world championship. At mid-season, the Pirates owned a 52-30 record, and a 44-29 run from July 17 until the end of September secured them a third consecutive NL East crown and a date with the Reds in the NLCS. Right-hander Steve Blass, fresh off his best season as a big-league starter, won as expected in Game One, and righties Nellie Briles, Bruce Kison and Dave Giusti limited Cincinnati to seven hits in a 3-2 victory for a 2-1 advantage. But even after the Reds tied the series at two games apiece, the Pirates appeared poised to advance to the Fall Classic against Oakland once taking a 3-2 lead into the bottom of ninth in Game Five.

Guisti started the final frame and jumped out to a 1-2 count against Hall of Fame catcher Johnny Bench. "To this day, I wish I would have never thrown the next pitch," the reliever explained. "It was a palm ball, and when I let it go I immediately thought I had him. It felt like the best palm ball I'd thrown all season. I guess Bench thought it was the best one, too."

The 14-time All-Star catcher tied the score, 3-3, with an opposite-field homer off Guisti. "I was more rattled on the mound after that home run than I ever was," Guisti added. "Before I knew it, there were two guys on base and no outs, and Bill Virdon was walking toward me with his hand out for the ball."

Virdon replaced Giusti with rotation member Bob Moose, a right-hander who had started 30 games and had made just one relief appearance all season. The manager's plan initially proved effective with Moose recording two quick outs, but with George Foster on third base and pinch-hitter Hal McRae at the plate, Moose hurled a wild pitch past backstop Manny Sanguillen.

"Whenever you lose, you always think about it, and a loss like that is going to linger for some time," Virdon

Bill Virdon – 43

said. "You think about what else you could have done, but eventually you come to realize that it's baseball. You have to learn to live with it.

"It's especially tough when I remember losing that one because of everything involved," Virdon continued. "Not only were we a better team in '72 than we were the year before, but we also lost an important player to a tragedy in the offseason. All of that made the next year very, very difficult."

Clemente, who collected his 3,000th hit in what proved to be his final regular season big-league at-bat, was killed in a plane crash on New Year's Eve 1972 while trying to deliver medical supplies and food to an earthquake-stricken Nicaragua. The Puerto Rican was immediately inducted into the Hall of Fame, the first Latino player to be enshrined in Cooperstown.

"Roberto was not a player you could try to replace, and I can't say that about too many players I've seen play the game. There have been a few, but not very many," Virdon said. "So not having Roberto made the '73 season a tough one from the first day until the last."

Getting fired in early September with the Pirates possessing a 67-69 record did not improve Virdon's situation in Pittsburgh. It is a move he disagreed with, and one that sent the "Quail" away from the Steel City for more than a decade. "At the time, I wasn't happy with the move to fire me. I didn't think it was right because we still had a chance to win. We still had a chance to go on and win our division and I wanted to have that chance."

"That was probably the dumbest thing," said Mazeroski, a utility player for the Pirates at the time. "It was not a good move. Murtaugh came in and he wasn't there all year. He didn't know what was happening. If you're with the team all year, you have an idea of who's doing what and how and who you would want to do what. If you're around it, you know. If you're not around it and then come in, you do bad things – and bad things happened."

Virdon returned to his wife, Shirley, and his three daughters residing at his Missouri homestead for the remainder of the season, but soon was hired by the Yankees. Although he was owner George Steinbrenner's second choice – the "Boss" first attempted to lure Dick Williams away from Oakland – Virdon's club completed the 1974 campaign in second in the AL East Division. The Yankees' skipper was named the Junior Circuit's best manager by *The Sporting News*, but the accolades did not matter when Virdon was replaced with Billy Martin on Aug. 1, 1975, although New York owned a 53-51 record.

Two weeks later, Houston hired Virdon to supplant Preston Gomez, and after seven years as an Astro, Virdon was replaced by Bob Lillis in August 1982. The former Bucco was honored with best manager honors from the United Press International in 1979, and by *The Sporting News* in 1980 when Houston qualified for its first postseason in franchise history.

Virdon's managing career continued for two more campaigns, 1983-84, with Montreal, until a 64-67 August mark led to his termination in favor of Jim Fanning. "I applied for a couple of jobs (in 1993 with Colorado and Florida) once expansion started," he said, "but those were opportunities that didn't work out. But I have no regrets.

"I was recognized in both leagues for doing a good job, and there are a lot of good managers who go their whole careers without that kind of recognition. Even though I wasn't managing anymore didn't mean I couldn't still work in the game. All I did was look for another job and I found one."

In 1986, Virdon joined Jim Leyland's first Pittsburgh coaching staff when the fiery manager took over a rebuilding process that eventually returned competitive baseball to Three Rivers Stadium after the sale of the ballclub and the infamous drug trial of 1985. Virdon then retreated to a roving position in the minors for three seasons, but he was present during the team's annual spring training session in Bradenton, Florida. In March 1991, a confrontation between Virdon and reining NL MVP Barry Bonds led to a full-scale scolding that made national news.

"That was an unfortunate situation – Barry was late getting to a drill and I said something to him about it, but what I heard come from him wasn't meant for me but someone else," Virdon explained. "But I didn't know that at the time, so after I said something back to him, Jim (Leyland) heard it, and he chimed in quite a bit from there."

The ballclub's financial heartbeat grew faint in the early 1990s. Bobby Bonilla's departure for the Mets via free agency following the '91 schedule, and the trade of southpaw starter John Smiley less than a month before Opening Day '92, were telltale signs of the decisions to follow.

"Those were tough years because of everything we had talent-wise, and everything we lost because of what the organization needed to do at the time," explained Virdon, who rejoined Leyland's staff in 1992 and stayed until 1996. "When you know you have something special but it has to go away because of the business, that's difficult to deal with – especially since we couldn't get past the NLCS for three straight years.

"Getting back into the competitive atmosphere in '97 with the Astros was a lot of fun for me because of what we dealt with in Pittsburgh after the three division titles with the players we had," Virdon said. "There have been a lot of changes in the game since I first came up (in 1955) as far as how many teams there are now and the salaries, but one thing will always be the same – you have to have the best team to win it all.

"We had that in 1960 and we had it in '71," Virdon added. "But we didn't in 1972, in 1980 (with Houston) or in the early '90s because we got beat before even getting to the World Series. That's what keeps you coming back, and that's why I decided long ago that I would keeping coming back as long as the game lets me."

5

Bill Mazeroski

Pirates Second Baseman, 1956-72

"Talk about modest. 'Maz' is a guy who gets embarrassed when you say he was a good player let alone the greatest second baseman ever to play in the majors."

— Nellie Briles

He has signed and signed and signed. And his hand, the same right paw that tossed a thousand baseballs more than a million times during a long big-league career in Pittsburgh, often crinkles with cramps.

"But that's never stopped the fans from asking," explained Hall of Famer Bill Mazeroski. "So what choice have I had? They cheered for me for a long time, so I think it's the least I can do.

"The people of Pittsburgh were behind me for all those years, and I think they felt a lot worse than me those years I didn't make it to the Hall of Fame. I know they were more upset because I got more letters for not making it than letters when I actually made it. Pittsburgh has always been behind me, and I've always felt proud because of it."

Autograph seekers have hunted Mazeroski throughout the decades that followed his historic, World Series-ending home run in 1960. Elvira Stogner of J. Paul Sports, a New York-based memorabilia company, said while Mazeroski owns several on-field records, he likely recorded another a few months after his August 2001 Cooperstown induction. "'Maz' signed for 18 straight hours for us after he got into the Hall of Fame, and that's the longest I've ever heard of," she said. "He started signing at 9 a.m. and wouldn't finish until he signed everything people sent in for his signature. He signed until early in the morning the next day. I don't know if it's an official record or not, but no one I know in this business has ever heard of a Hall of Famer signing for that long in one sitting.

"He didn't have to do that, but he did for his fans. He said that if they went to the trouble to send the balls, the photos and the posters to us, that he owed it to them. You don't hear that from everyone. There are a lot of former athletes who are happy to sign, but only during the hours they're paid for. But that's not 'Maz'."

Mazeroski, a long-time Pirates Fantasy Camp favorite who also served as a special spring training instructor for several seasons, did admit to tiring of the constant requests. "After I got into the Hall of Fame, it was crazy. It seemed like I was at a different (memorabilia) show every weekend and I got tired of the travel. I never get tired of meeting the fans and hearing their memories, but I've turned most of the shows down since then.

"When I've been at spring training, I've spent more time on the field than anything else," the West Virginia native said. "I can hide out there. As soon as I get close to where the fans are, I could be there for hours and that's when my hand cramps up. I think it's the baseballs – I get tennis elbow from holding too many baseballs. You have to hold it right and then you got to roll it right. The tennis elbow hurts my golf game."

Bill Mazeroski – 51

The demand, Mazeroski acknowledged, is understood because Hall of Fame status followed a modest, small town upbringing. He was an All-Ohio hoopster and was offered several athletic scholarships while attending Warren Consolidated High School in tiny Tiltonsville, Ohio, but when the time arrived to decide, Mazeroski chose baseball over basketball. "I liked basketball a lot, but baseball was more in my blood than basketball. I got scholarship offers from a lot of colleges, but basketball involved school. Baseball didn't involve school. That probably had a lot to do with it. Plus, I could play baseball and get paid for it? Wow. It wasn't that tough of a decision."

Two minor-league seasons after signing with Pittsburgh in 1954, Mazeroski made his debut at second for the Pirates and took over for Johnny O'Brien, playing 81 games at 19 years old and hitting .243 with 62 hits in 255 plate appearances. Over the next 16 years, Mazeroski became known as "No Touch" while establishing major-league records for most career double plays by a second baseman with 1,706 and the most twin killings at the position in a single season with 161 in 1966. He paced all National League second baseman in completed double plays for eight consecutive seasons (1960-67) and in assists for nine seasons, both of which were immediately considered untouchable Senior Circuit standards upon his retirement after the 1972 campaign.

His eight Gold Gloves and 10 All-Star selections mean less, Mazeroski insisted, than the three division titles, the pair of National League pennants and the two World Series championships he helped Pittsburgh claim while in uniform. "You play to win games, and in the end, you hope you can take home a trophy," he said. "I never played as an individual because that kind of stuff didn't get the job done. It's a tough game, and it's a long season.

If you don't go out there as a team, you can't expect to win as a team. Everyone has to get their job done, and when that happens, usually good things follow.

"I'm proud of the Gold Gloves, especially after playing at Forbes Field all those years. That field wasn't as smooth as they are today, that's for sure," continued Mazeroski, whose No. 9 was retired by the Pittsburgh organization in 1987. "And I considered being invited to the All-Star games a big honor, but I was only as good as the other guys on the team. We won a few times because we played as a team, not because I played second base or because Roberto (Clemente) was in right, or because of who was on the mound. It was because 25 guys got their jobs done."

Mazeroski's propensity to distribute credit to his teammates, and away from himself, has been appreciated by those with whom he suited up during his career. "Talk about modest," said former Pirate Nellie Briles, a Pittsburgh pitcher from 1971-1973. "'Maz' is a guy who gets embarrassed when you say that he was good let alone the greatest second baseman ever to play in the majors.

"I've known 'Maz' for a long, long time, and never has he admitted that he played a key role on the teams that had success, and it's not that he was just trying to say the right thing like some guys do. I think he truly believes he was just one man on a team, and when you say something like, 'But you hardly ever made errors,' he'd answer, 'I wasn't being paid to make errors so I did my best not to do that.'"

Manny Sanguillen, a teammate of Mazeroski's from 1967-72, referred to the infielder as the veteran mentor he needed upon his arrival from Panama. "I remember when I came to the big leagues and my locker was next to his. He looked at me in the chair and said, 'Manny, I want you to stay there. So many guys come in and go.

You stay there, Manny. You don't want to go anywhere else.'

"'Maz' used to talk to me a lot because he believed in me, and I needed someone to believe in me when I first came to this country to play in the major leagues. I was happy for him when he finally got into the Hall of Fame, but 'Maz' is one of my favorites because he was one of the people who took care of me. For that, I will always consider him a great, great man."

Mazeroski joined 11 former Pirates in the National Baseball Hall of Fame on Aug. 5, 2001, and although a planned 12-page presentation was displaced by a brief, emotional speech, his plaque placed on display forever the grace and grit with which he played:

A defensive wizard whose hard-nosed hustle and quiet work ethic helped lead the Pirates to three division titles, two pennants and a pair of World Championships. An eight-time Gold Glove winner and a 10-time All-Star renowned for his lightning-quick pivot on the double play, turning 100 or more in 11 consecutive years. His 1,706 career twin killings is a record among middle infielders. Also ranks in the top 10 among second baseman in assists, putouts and games played. His dramatic home run in Game Seven at Forbes Field propelled the Pirates to the 1960 World Championship.

"Even when I was practicing that speech, I knew I wouldn't make it through the whole thing. I had tears in my eyes when I was writing it," Mazeroski said. "I knew I wouldn't get it in. But there were 12 pages, and they're locked up in a safe somewhere. The print was pretty big so I could see it, but there was a lot there.

"I said what I needed to say – what I really wanted to say – that defense deserved a place in the Hall of Fame. I think they've concentrated on the home runs and offense too much so when I went up there I made sure I told the folks in the crowd that defense was just as important as offense. As far as thanking everyone I wanted to thank, I've done a pretty good job of that after that weekend. There were a lot of names on those pages."

His wait for induction extended nearly three decades. Once Mazeroski was eligible in 1978, voting members of the Baseball Writers Association of America passed him over for 15 years. His candidacy was then handed over to the Hall's Veteran's Committee in 1993, but while the panel paved way for Phil Rizzuto, Richie Ashburn, Vic Willis, Jim Bunning, Nellie Fox, George Davis, Larry Doby, Orlando Cepeda and Bid McPhee, Mazeroski remained patient without prejudice. "All those guys were great players, so why would I have any problems with them being put into the Hall of Fame? They deserved it, that's for sure.

"You know, I really never waited for it because I truly never expected it," he said. "The media made me feel like I was on pins and needles every time, and that was never the case because I never thought it would ever happen. I never thought defense would get that much respect."

Mazeroski's run production statistics – a .260 career batting average, 128 home runs and 853 plated runs – were the numbers that provoked the defense against his Hall entrance. Not until former Pittsburgh GM and Veteran's panel member Joe L. Brown contacted him the afternoon of March 6, 2001, at McKechnie Field in Bradenton, Florida, did Mazeroski believe his golden glove meant as much to the game's history as baseball's legendary sluggers.

"I worked hard on my defense because that's how I helped the team the most," Mazeroski said. "It was different when I played – when you come through baseball these days there are instructors at each level to teach you everything. When I came up, you had to learn every-

Bill Mazeroski – 55

thing yourself through trial and error. I learned how to do the double play through trial and error. I had no one to show me. If you asked the other older players, they wouldn't show you because you were going to be taking their job sooner or later. They didn't bother with you, and you learned on your own. Now, you have an instructor at every level and probably two instructors for each position during spring training. It's a lot different these days. It's a lot different.

"My job wasn't to hit the home runs, it was to be the best middle infielder I could be, so that's what I tried to do. I can tell you now that I didn't really concentrate at the plate. I did when there were runners on base, but usually I would go up there, look for the ball and just swing at it. With men on base, I concentrated on the ball more and hit it hard."

That adopted approach led to the legend born with Mazeroski's final swing of the '60 season. The infielder batted .320 in the Fall Classic, and his two-run homer in Game One helped hand the Pirates a 6-4 victory. But never had a man ended a World Series with a walk-off home run until Mazeroski deposited Ralph Terry's 1-0 offering over Yogi Berra's head and beyond the left field wall of Forbes Field to propel the "Battlin' Buccos" to a 10-9, Game Seven triumph.

"I went up only thinking about getting on base," Mazeroski explained. "We lost the lead in the top half of the inning, but there was no way we were quitting. I went up to the plate just thinking about getting on base, and Ralph put one over the plate and I hit it pretty hard. I've heard it was supposed to be a slider, but it sure looked like a fastball to me.

"But that was just one swing that season. So many guys did the same kind of thing that year. Hal Smith hit the three-runner in the eighth that gave us the big lead that we lost, and that's just another example of the kind of things we did that season to get into the position to win the World Series. I think it was (outfielder) Gino Cimoli who came up with that saying – the "Battlin' Buccos" – and that was because we never quit.

"We were the Battlin' Buccos' before I hit that homer. I think we got that nickname because we would come back game after game. From the seventh inning and on, we would win. We'd be down and then we'd come back. I was just one guy. I didn't consider myself anything special. The homer? Yeah, I hit the homer, but that wasn't why we won the World Series."

Yankees fans and members of the New York's '60 ballclub have disagreed, and Mazeroski said he has encountered their collective disappointment on several occasions since the Pirates claimed their fourth big-league title. "A lot of Yankee fans haven't gotten over it from what I can tell. They remember that one loss, I think, more than all the wins they have had," he explained. "Ralph Terry has always been nice to me, but, of course, I never bring the homer up around him. He does, sooner or later, but that's not my style.

"Now, teaching Yogi Berra when over is over? Yeah, that's something I still enjoy. He saw it better than anyone when it went over his head, and I still agitate him a little about that when I see him. He keeps saying it scrapped the wall and I've always told him that it cleared by 20 feet. He still hasn't gotten over it."

Mazeroski, who chose a private retreat to Schenley Park with his wife, Milene, after Game Seven instead of participating in the raucous parade in downtown Pittsburgh, certainly did not disappear following the '60 season. He was bestowed several mid-season and post-season honors in the 1960s, and in 1968 he appeared with other Pirates in the movie version of, "The Odd Couple."

In 1969, however, leg injuries limited his play to 67 appearances, and skipper Danny Murtaugh was forced to depend on young prospect Dave Cash. By the time Pittsburgh found itself back in the World Series in 1971 against Baltimore, Mazeroski was 34 years old and had collected his 2,000th hit, but his role altered from everyday player to coming off the Pirates' bench. During Pittsburgh's win over San Francisco in the '71 National League Championship Series, "Maz" was 1-for-1, and in the Series victory, he was hitless in one trip to the plate.

"It seemed like every time I did get out there I'd pull a muscle. I was always getting hurt so I knew it was time. We didn't take care of our bodies like they do nowadays," he said. "We didn't stay in shape year-round; we didn't work out; we didn't lift (weights); we didn't stretch. All of a sudden, I wasn't a 19-year-old rookie anymore. There was a new crop of younger players coming up and taking over at a lot of positions, so I could read the writing on the wall.

"Back then, we just went to the spring training, got in shape and played. That shortened your career if you had a body like mine. I was thick and not fit, so if I would do something to stretch, something would pop. I'd pull a groin or pull this or that, and when it didn't get any better, I knew it was time to quit."

When Mazeroski returned to the Pittsburgh roster in 1972, the infielder was confident it would be his final season. He appeared in 34 games, collected just 12 hits and scored three runs as Pittsburgh won a third straight division title, and he retired soon after Cincinnati's "Big Red Machine" defeated the Pirates in the NLCS. "You could tell the Pirates were gearing up for a good decade of baseball, and they played some great baseball in the 1970s. As much as I would have liked to be part of that, it was definitely time for me to get out of the way.

"I couldn't do the things I had to do on the field, and that stayed on my mind," Mazeroski said. "The toughest part was probably not going to spring training the next year – just getting packed and ready to go. When January came, I always went down early to work out. That was the hard part. When that next January rolled around and I didn't have anywhere to go, that's when I really realized that it was over and that it was time to sit back and relax like an old timer should."

6

Steve Blass
Pirates Starter, 1964, '66-74

"Steve Blass was able to do anything. When he was pitching, the ball would move everywhere. He was smart, too. He would be able to see what the hitters did and be able to locate the ball at the right place at the proper time. And then it was all gone, just like that."

— Manny Sanguillen

His head sunk, and everyone at McKechnie Field – on the field, in both dugouts and in the stands – dropped their heads and covered their eyes as Steve Blass walked off the mound, and away from the game of baseball, for the final time.

Blass, a World Series hero to Pirates fans after tossing a pair of complete-game victories against Baltimore in the 1971 Fall Classic, lost his ability to throw consistent strikes one year following his best major-league season ever. In 1972, he represented the Pirates at the All-Star Game and finished second in the National League Cy Young balloting after posting a 19-8 record with a 2.49 earned-run average. In 1973, however, Blass was just 3-9 in 18 starts, and his ERA ballooned to 9.85 as he walked 84 batters and beaned 12 in 89 innings.

His '74 season began in Pittsburgh, but after his lone, five-inning outing, Pirates general manager Joe L. Brown demoted Blass to the Charleston (West Virginia) Charlies. "When I went to the minor leagues, there was sort of like a circus or carnival atmosphere," Blass explained. "The fans were saying, 'Let's go see the guy who can't throw strikes anymore.' It was tough, especially after winning the World Series, but I had to try, and when the decision was made that I couldn't try anymore on the major-league level, I understood it."

Although he issued 93 free passes in 57 frames for the Charlies, Blass was invited to spring training in 1975 for one last chance. Manager Danny Murtaugh sent the right-hander to the mound against the White Sox, and Blass tossed 6 2/3 innings and surrendered 13 runs and walked 17 batters. The final frame was the most painful with Blass issuing eight free passes and forcing in six runs.

"We all had our heads down when he walked off the field because we didn't want to see it," recalled former Pittsburgh catcher Manny Sanguillen. "Everyone in the stadium that day knew it was the end, and no one wanted to admit it. Steve Blass was too good, but whatever it was, it finally got him."

In the years that followed his release from the Pirates at 33 years old, the mysterious ailment became known across baseball as the "Steve Blass Syndrome," and the pitcher's two-season struggle to locate the strike zone forced him far away from the diamond dreams Blass owned since his childhood in Canaan, Connecticut. Less than a month after his last rotation turn, Blass became a traveling class ring salesman with Jostens, Inc. Once he completed a training session in Owatonna, Minnesota, he adopted a 55-school territory in Western

Steve Blass – 63

Pennsylvania.

"What happened with my control didn't sour me on the game itself, but it was enough to make me want to get completely away from it. I wanted some removal at that point just because it just wasn't that much fun," Blass explained. "The toughest part of the ring business was convincing these high school kids that, after being a major league ballplayer, I was serious about the business. Along with selling the product line, I had to sell myself not as a so-called celebrity who was just going to walk in and make sales, but as someone who actually knew the business and the product.

"I didn't make a hell of a lot of money, but I made enough to support my wife (Karen) and two sons (David and Christopher) and that's all I cared about after I was out of baseball. I only lasted eight years, though, because I got tired of putting my livelihood in the hands of 16- or 17-year-old youngsters who might make a decision to go with my company or another company based on a whim or how somebody dressed."

Although Blass was constantly confronted with questions about his abbreviated playing career, it was not until an opportunity arose with a Pittsburgh beer distributor that the former hurler started talking baseball once again. Blass visited Pittsburgh area taverns, showed highlight reels on the Pirates' and Steelers' championship seasons in the 1970s, and raffled off T-shirts and ballcaps.

From 1975 to 1985, he also orchestrated sessions of the "Steve Blass Baseball Camp" at The Kiski School in Saltsburg, Pennsylvania, and at The Linsly School in Wheeling, West Virginia. "I was very proud of those camps because I've always believed kids are very important to the game. If children fall in love with baseball at an early age, they'll either be players or fans when they grow up, and the game needs both.

"Meeting the fans at the bars was good for me because a lot of those people at those locations were sports fans and they remembered the good years," Blass continued. "I probably made about 300-to-400 of those appearances – lugging a projector and a screen on winter nights – but it was fun because I have always loved the game of baseball and I missed it very much. I never wanted to leave the game."

Blass visited eye doctors, hypnotists, and psychotherapists. He tried meditation and sought advice from coaches, teammates, and opponents – even fans – before that final day as a major-league pitcher. Possibilities were publicly investigated and discussed by the print media and radio talk show hosts. Was he unable to throw strikes because of a fear of beaning a batter? Did he, for whatever reason, lose confidence in his abilities? Or, did the sudden and tragic death of teammate Roberto Clemente on New Year's Eve 1972 shake his faith in fate and erect a permanent block on the strike zone?

"I still have no clue what caused it to happen," Blass said. "There wasn't a day when I suddenly couldn't throw strikes. It was a gradual thing that got worse as time passed. I was 3-3 early in 1973, and it never got any better after that.

"I was out there dangling because nothing I tried helped a bit. I was out in my backyard at 4 a.m. a lot of nights with me, myself and I, just wondering why it was going on. I have never come up with the answer and I was all over the place trying to find out why this wasn't working anymore. I tried everything. I wanted to find out if I could get it back, first of all. If I couldn't, I wanted to be satisfied that I wouldn't be sitting on a front porch in a rocking chair at the age of 75 thinking, 'Boy, I wish I had tried that.'"

His absence from Pittsburgh's rotation did not stop

Murtaugh's teams from collecting division titles in 1974-75, but his presence and experience would have been appreciated once Chuck Tanner was traded for and hired following Murtaugh's death on Dec. 2, 1976. Murtaugh's final season saw Pittsburgh finish second to Philadelphia, and Tanner's first two ballclubs also completed regular seasons as the National League East's runner-up.

"Who knows what kind of pitcher he would have been by the time I came to the Pirates," Tanner said. "But I can tell you that our pitchers would have benefited from just being around him because of how smart he was on the mound, and it wouldn't have mattered if he was still a starter at that point or if he needed to be a reliever instead.

"I was still managing the White Sox in the American League when he had his best year, and then when he was going through all his problems with his control. But everyone in baseball knew who Steve Blass was, and about his struggles. If I could have helped him, I would have, but no one knew why, all of a sudden, Steve couldn't throw strikes anymore."

Although Murtaugh often positioned Sanguillen in right field in 1973, the Panama native was behind the plate for the two gems Blass tossed in the '71 Series and throughout the 1972 campaign. "Steve Blass was able to do anything. When he was pitching, the ball would move everywhere. He was smart, too. He would be able to see what the hitters do and be able to locate the ball at the right place at the proper time. And then it was gone, just like that.

"When he was at his best, it was because of his slider," Sanguillen explained. "He could throw it at any time, and he could throw it on the inside or the outside of the plate. But then he lost the slider, and his curve, and his fastball, and I don't know why. We all tried to figure it out."

Nellie Briles was acquired by Pittsburgh prior to the '71 schedule and played a key role in the club's World Series triumph, pitching the Pirates to a complete-game, 4-0 victory in Game Five. He remained with Pittsburgh until the right-hander was traded to Kansas City following the 1973 season, and was exposed to the trials and tribulations of the "Syndrome."

"Steve Blass was a guy who was a control artist and relied a lot on control and then, all of the sudden, he couldn't throw a ball in the batting cage. Your heart bled for him," he said. "We all had that feeling in the pits of our stomachs because he still had many years to pitch.

"We'd watch him to see if we could figure it out, and no one could. His motion was unusual, but he looked like the same pitcher, mechanics-wise, as he did when he was one of the best in '72, so there was nothing to say about his delivery. And then he went to every professional there was at the time, and they couldn't figure it out. It was a mystery then, and it's remained a mystery to all of us."

Hall of Famer Bill Mazeroski, a teammate of Blass' from 1964-1972, believed the right-hander would become one of baseball's best after witnessing firsthand the hurler's dominance. "There were times when he was doing whatever he wanted. He owned the hitters on some days.

"But then something went wrong in his head and it's been something no one has been able to figure out. It had to be something in his head because he wasn't injured or anything like that."

Other pitchers who have lost control on the pitching mound since Blass confused baseball with the "Syndrome" include Detroit closer Kevin "Hot Sauce" Saucier (1982), Philadelphia pitchers Joe Cowley (1987) and Bruce Ruffin (1988), Kansas City's Mark Davis

(1990), Atlanta's Mark Wohlers (1998) and the Cardinals' Rick Ankeil (2000).

Blass has chosen not to communicate with those who have been infected with the disease that bears his name because, he explained, he never solved the mystery. "I didn't get it back, so why would anyone want to talk to me? Whatever it was, it ruined my career, but those guys had a chance to get it back. Talking to me might have only made them lose it forever because I did," he said. "Now, I have talked to Mark Wohlers, but that was because he sought me out. All I did was encourage him to try absolutely everything because there might be one little sliver or one little item that he may stumble on while trying everything.

"During those two years, I honestly felt it would take just one little thing for me to get it back, but I never found that one little thing," Blass explained. "You know, I don't hate that this thing has my name on it and every time a guy suddenly can't find the zone, but it's not my favorite thing. I can understand the reasons. I did go from finishing fourth in the Cy Young voting to being out of baseball two years later without having a sore arm. But I've moved passed it, and I think I've done OK."

Blass' career path did finally lead him back to the ballpark. In 1983, he joined the legendary Bob Prince as a guest color analyst and he continued making frequent appearances until joining the Pirates' broadcast team in 1986. "I liked the idea of broadcasting even when I was playing and I used to talk to (former Pirates pitcher and broadcaster) Nellie King about it. But it was tossed to the background because of all the negatives the last couple of years of playing baseball.

"But then, in the early '80s, the Pirates were looking for a fourth announcer. They had (Jim) Rooker, Lanny (Frattare) and John Sanders, and they were looking for a fourth guy so they could have two on TV and two on radio," Blass explained. "They brought in several people just to sit in on a game, and I was one of those guys invited. I'm guessing they liked what they heard."

"What I admire the most about Steve," Briles said, "is that he did not let the loss of his baseball career ruin his life. He has gone on to establish another career and does a great job. There are so many people who can't deal with having that type of episode in their life, but he picked himself up, turned the page and has really made something of his life. I think that says a lot about the person Steve Blass has become."

Although his broadcasting career has lasted longer than his playing days, Blass said he was blessed to be involved with the game, especially in the western Pennsylvania region. "Pittsburgh is a great baseball town, and the fans here want winning teams on a consistent basis or they get annoyed," he said. "And there's nothing wrong with that, nothing at all.

"And I've been a lucky man, being able to be a player and then to be in the broadcast booth for so many years. Jim Leyland said it best, I think. He said that I've had the two best jobs in baseball – a starting pitcher who only has to work once a week, and then a professional bullcrapper."

Blass also found peace with his "Syndrome" in 2000 after meeting Richard Crowley, a psychologist from New Mexico who is known as a "performance coach." During a chance meeting in Bradenton during the ballclub's annual spring session, Crowley became intrigued with the pitcher's problems. "We began talking about it, and although I've always been very polite when hearing all the remedies and suggestions I've been offered, I thought he had a very interesting way of presenting his theories. That's why I decided to work with him for a little bit, and I've been grateful ever since.

"As a result, the next spring I started meeting with (minor-league manager) Trent Jewett at Pirate City at 7 a.m. I wanted to see if I could recapture the joy of throwing. I have always loved just throwing the baseball, and I wanted to see if I could get that back. It was a wonderful, wonderful time in my life to go back and throw the ball and throw it with conviction again."

In nine major-league seasons, Blass compiled a .575 winning percentage (103-76) and a career 3.63 ERA. Since adopting broadcasting as his primary vocation he has earned the reputation as one of the best in baseball, and was honored by the Pittsburgh organization in 2002 with the "Pride of the Pirates" award.

"Because I was able to go out and throw a baseball again, I can say that I'm satisfied with how everything has worked out. When I've thought about it all, I've realized that I've done all the things I dreamed about times-10. I pitched in an All-Star Game. I pitched in the World Series, and I won 100 games," Blass said. "I had teammates who were unbelievable, both when I was going good and when I was going bad, and they probably had no idea how much they helped me get through those two unbelievable bad years. Nobody stood taller than Dave Guisti, Willie Stargell and Bill Mazeroski when I was going bad.

"I can say that I've done everything I wanted to. I had the hat trick – I've had one team, one house and one wife."

7

Manny Sanguillen
Pirates Catcher, 1967, 69-76, 78-80

"Everyone has silly stories about Manny, but he called a pretty damn good game and he knew the league. I was real fortunate to have him there at that particular time in my career even though I really couldn't understand much of what he said to me on the mound. Steve Blass is right when he says Manny is the only guy who has been in the country for 35 years and his English has gotten worse."

— Kent Tekulve

He smiled constantly and swung consistently at pitches nowhere near the strike zone. But while nearly 1,600 baseball players have donned Pittsburgh uniforms during the Pirates' first 117 seasons, not many have become as famous as Manny Sanguillen. Twelve former Pirates have earned spots in the National Baseball Hall of Fame, yet it was Sanguillen who attracted the attention of a Pulitzer Prize-winning novelist, and he was immortalized with his very own Bobblehead doll in 2003.

"And I am the only one of the alumni (bobbleheads)

who isn't in the Hall of Fame," Sanguillen said while flashing his infectious, trademark, gapped-tooth smile. "I told 'Maz' that he wasted all that time waiting to get into the Hall when all he had to do was get a bobblehead. That's my trophy. That's all I need."

During his 13-season career, Sanguillen earned reputations for both his broad grin and his free-swinging approach at the plate. In no one season did the Panama native collect more than 48 free passes. "I don't think too many pitchers liked it when I smiled because a lot of them threw the ball right at me," he said. "They thought I was making fun of them or showing them up, but I was just having fun playing baseball.

"I didn't feel I had time to wait for the ball to come to me," he explained. "I felt I had to do good right away because I was getting started late. When I came up to the big leagues, I wasn't 23 years old like everyone thinks. I was 27."

Pittsburgh's 1971 media guide states that Sanguillen was born on March 21, 1944 in Colon, Panama, but the former backstop corrected the publication. "My birthday is on March 21, that's true. And I was born in Colon, but it was in 1940," he said. "I didn't lie about my age. That's what it had on my first contract, and I didn't know anything about baseball at the time."

Several of Sanguillen's former teammates were not surprised to find out "Sangy" was older than advertised. "No kidding," Hall of Famer Bill Mazeroski said. "I don't know if he was trying to fool anyone, but he wasn't fooling me. I think we all knew he was older than what it said on his baseball card."

"Only four?" former closer Kent Tekulve said. "The only thing that I'm surprised about is that it wasn't more than four. I'm pretty sure everyone had an idea he was older than what was said, but there were a lot of guys like that."

Manny Sanguillen – 73

"Some of us thought 'Sangy' was born at the turn of the century – the 20th Century," joked former starter Nellie Briles. "I guess four years isn't that bad, but there were a lot of us who believed it was much more than that."

Sanguillen ran track and played basketball and soccer for Abel Bravo High School before working as a fisherman and moonlighting as a boxer in his native Panama. Not until he was in his 20s did he first set foot on a baseball field. "First, I played softball and people wanted me to go to Panama City to play in a baseball league, so I went," he explained. "Then a scout (Howie Haak) with the Pirates saw me play and wanted me to go to the United States to play."

His professional career began in 1965 with Class A Batavia, and two years later he appeared in 30 games for the Pirates. After playing a full season in 1968 for Class AAA Columbus, Sanguillen officially arrived to majors. "Things weren't easy for me then because I was a Latino catcher and there weren't many of them. I was one of the first. It is true that I learned a lot about baseball from watching TV. I watched what the catchers did and how the pitchers threw the games.

"Even one of the team's scouts (Harding Peterson) said I would never play in the big leagues because I didn't go to an American high school. He said I would never learn how to call pitches.

"Peterson? He later became the general manager (in 1979) and he told me he was wrong about me," Sanguillen continued. "I just smiled at him and he gave me a big hug. I knew I was blessed to play baseball, and I was very proud to be a Pirate. That's why I always smiled."

Following his 1980 season with Pittsburgh, Sanguillen decided to retire instead of reporting to Cleveland's spring complex following a trade that also involved starter Bert Blyleven. His career average stood at .296, and in 13 years Sanguillen helped the Pirates capture six division titles and a pair of world championships.

"And I was 40 years old, not 36, don't forget," the former backstop said. "My plan the whole time was if I was going to hit the ball and help the team win, I had to swing the bat. I loved my teammates and the best way to help them was to hit a lot.

"When I was traded to the Indians, I had played enough and didn't want to go. I still loved going to the ballpark and playing baseball, but I couldn't play the same like when I was younger."

Briles, who was a starter in Pittsburgh from 1970-73, said he quickly came to appreciate Sanguillen's abilities behind the plate and in the batter's box. "You know, Manny may have looked like a nice guy with that big, split-toothed smile, but he was competitive as could be and as fine an athletic catcher I ever threw to. He could move. He had a great arm, and I loved throwing to him because I was a low-ball pitcher and he got down on his haunches with his fanny on the ground.

"Now, when he was hitting, you never knew what he was going to swing at, that's for sure," Briles continued. "He always made contact, and his career average is a testament to that, but were there ever some interesting moments when he was at bat."

Michael Chabon, a former University of Pittsburgh graduate student who was awarded the Pulitzer Prize for fiction in 2001 for *The Amazing Adventures of Kavalier and Clay*, praised Sanguillen during a talk at the Carnegie Music Hall. The novelist, who also mentioned the Pirates in *Mysteries of Pittsburgh* and *Wonder Boys*, said:

"Roberto Clemente was the first person I loved ever to go and die on me. The city of Pittsburgh, however, did not abandon its place in my imagination, because there was this catcher the Pirates had. Clemente had overshadowed him, at first, in my imagination. But after Clemente was gone I started to notice how this guy, Manny Sanguillen, seemed to have discovered some inexhaustible source of joy, a source that at the time I took to be baseball itself. He was a free-swinging, dirty-uniform kind of guy, and he was always smiling. It made *you* smile, just to look at him. I don't know if you remember or not, but the middle of the 1970s was not exactly the most joyous of times. The whole world seemed to have turned into a Robert Altman movie. Jarring and sour, and crazy, and colored in a palette that I believe drove my entire generation mildly insane. This malaise—that's how Jimmy Carter later styled it—had invaded baseball, too. Among the bizarre, misbegotten adventures of that time, along with the bombing of Cambodia, wife swapping, and necklaces for men, was the invention of the designated hitter rule—and just think of the Houston Astros uniforms from that time! They could be seen from outer space! But there was no *malaise* for Manny Sanguillen. He went on swinging at bad balls, and tearing around the bases faster than any catcher had a right to do. This smiling and delightful man, I supposed, also made his home in that fabulous city that had spawned Roberto Clemente: Pittsburgh, Pennsylvania."

"People liked me because of the way I played the game," Sanguillen said. "They saw that I had fun out there, and that I would sacrifice everything to be able to win more than lose."

Sanguillen's smirk, though, did vanish for an extended period of time as he joined the entire baseball community in mourning the tragic and sudden death of Clemente. The plane crash that claimed the lives of Clemente and four other men set to deliver assistance to Nicaragua following a deadly earthquake after a disappointing end to the Pirates' 1972 season.

Pittsburgh claimed its fourth world title in '71 in a seven-game World Series in which Clemente shined in front of a national audience. In 29 turns at the plate, the Hall of Fame right fielder collected 12 hits – including two home runs – and was named the Fall Classic's most valuable player. "That World Series meant so much to him," Sanguillen recalled. "For so long he was discriminated against, and people said things like he was a hypochondriac and that he didn't really care about winning. None of that was true.

"Clemente was the type of person you wanted to listen to if you wanted to learn baseball and wanted to win. Do you know how people treated him because he was Latino? That was really bad at that time. Little did they all know he was going to lead them to the promised land.

"Roberto was the best teacher. We loved him because he loved the Pittsburgh Pirates, and he loved us. When a rookie would get to the big leagues, Roberto would sit with him and welcome him. He said to me, 'You're here like me. Don't worry that I'm who I am. You're just like me. You're a big leaguer. If there is anything I can help you with let me know. If you need a bat or glove, you come to me. Whatever you need.' He made you feel like you were home. Clemente was really 20 years ahead of the game."

It was during that magical '71 season when Sanguillen unseated perennial All-Star Johnny Bench as the NL starting catcher, and at year's end the Panamanian owned the third-best batting average in the Senior Circuit (.319).

Although Clemente collected his 3,000th base hit on the final day of the '72 campaign, the Pirates failed in their attempt to repeat as world champions. Pittsburgh claimed the National League East Division with a 96-59 record but lost in the championship series to Cincinnati three games to two thanks to a two-run Reds rally in the bottom of the ninth.

Ten weeks later, on New Year's Eve, the 38-year-old Clemente chartered a DC-7. The flight, scheduled to depart Luis Munoz Marin International Airport in San Juan, Puerto Rico, was delayed for more than 15 hours because of mechanical problems with the engines. The plane's first attempt to take off, in fact, was unsuccessful.

Clemente, said Sanguillen, was not deterred by the delays only because of circulated rumors that looters had seized many of the valuables sent by the outfielder in previous flights. "He wanted to make sure the people who needed the supplies the most would get them this time, and it did not matter to him that it was New Year's Eve."

The DC-7 lifted off loaded with approximately eight tons of supplies at around 8:45 p.m., but two minutes later the craft crashed into the Atlantic Ocean in 150 feet of shark-infested waters. Only the body of the pilot, Jerry Hill, was discovered. "The plane was overloaded because he was trying to take more than usual because he heard they were stealing. They overloaded it so much with baggage that Clemente had to take his jacket off and lay on the floor of the plane. I really think he was sleeping when everything happened. At least I hope he was."

Soon after hearing of the fatal crash, Sanguillen raced to the shoreline where he could see the helicopters' searchlights over the point of impact. Sanguillen, grief stricken and reacting in panic, borrowed a boat and made his way to the area. For more than an hour, Sanguillen dived repeatedly into the dark, dangerous ocean tide.

"I didn't care about the sharks – I knew about the sharks, but I didn't care, I just wanted to see Roberto again," Sanguillen remembered. "I kept thinking about the first time I ever saw him. It was on an airplane, and people kept coming up to him because they wanted to meet him because he was so great. They wanted to touch him, his greatness. That's what I wanted the most on the night he died.

"The thing about me and Clemente was that he didn't just teach me about baseball, he taught me about life. And I wasn't ready to do it by myself. Then, when we started the next season, his family asked me if I would play right field for him. I didn't want to because I knew I didn't belong there. No one did, but we needed someone and I was the one."

Sanguillen was positioned in Clemente's place on Opening Day 1973, and 58 more times during the course of the season, but the loss of the All-Star outfielder coupled with starter Steve Blass' sudden loss of control resulted in an 80-82 record and a third-place finish in the NL East Division. "Nothing was right about that season," Briles explained. "There was a time when we all thought we could play through all of the grief, but it caught up with us for whatever reasons.

"Roberto truly had an impact on everyone in our clubhouse, and when he wasn't there anymore there was no one who was ready to step in and take over his role. I felt bad for Manny, I really did. A lot of people looked to him for something – anything, really. I think then we all realized there was only one Roberto Clemente, and I think baseball has realized that in the years that have followed. There's no way to explain it."

In 1974, Sanguillen was back behind the plate, his smile gleamed again, and the Pirates claimed yet another division title after completing an 88-74 regular season. A 5.14 earned-run average and a team batting average of .194, however, spelled doom for Pittsburgh in the Dodgers' three-game-to-one NLCS triumph.

"Manny was the first guy I leaned on when I first came up to the majors in '74," explained Tekulve. "Everyone has silly stories about Manny, but he called a pretty damn good game, and he knew the league. I was real fortunate to have him there at that particular time in my career even though I really couldn't understand much of what he said to me on the mound. Steve Blass is right when he says Manny is the only guy who has been in the country for 35 years and his English has gotten worse.

"But he's the guy who made me realize as a rookie that the mound was the same distance in the big leagues as it was in the minors. And I was just like Manny – we were guys no one thought would ever make it to the big leagues, and when I got there, I was nervous. He took it right out of me."

But Sanguillen achieved his primary goal and established himself as one of the best offensive catchers in the NL. It was during the 1975 season when he was at his best, ranking third once again in batting with a .328 average. The backstop also recorded a .391 on-base percentage thanks to a career-high 48 walks as the Pirates captured the East Division yet again – and lost to Cincinnati's "Big Red Machine" once again in the NLCS. His success at the plate continued in 1976 with a .290 pace, but on Nov. 5, "Sangy" was sent to Oakland for their manager, Chuck Tanner, and $100,000 in cash.

"I didn't laugh at all about it because I didn't want to leave the Pirates," Sanguillen recalled. "I had made Pittsburgh home, and the last thing I expected was to be traded for anyone. But I told myself that if the Pirates didn't want me anymore, I'd go wherever they wanted me. I didn't say anything to anyone. I guess it was in God's plan."

He spent just one season with the A's, serving as Oakland's designated hitter while also playing first base, in the outfield and behind the dish. He batted .275

in 571 at-bats for a last-place, 63-98 ballclub and was left with one season highlight – on two consecutive days he broke up no-hit bids against the Yankees' Mike Torrez and Baltimore's Jim Palmer on Aug. 11-12. "I wasn't used to that. I was always playing for first place, but we had no chance when I was in Oakland," Sanguillen explained. "The best thing was to get traded back to Pittsburgh.

"When I got traded back (for Mike Edwards, Miquel Dilone and Elias Sosa), I thanked God. In my head, I always knew I was a Pittsburgh Pirate. I was a 'Bucco,' and it was a blessing."

But Tanner already possessed his full-time catcher in Ed Ott so Sanguillen's role significantly diminished upon his return. In 1978, in fact, Sanguillen appeared in only 18 games at catcher, but he did spell Stargell at first base on 40 occasions and batted .264. His playing time decreased during the Pirates' world championship season in 1979 as he was used primarily as a pinch hitter off Tanner's bench.

Although he appeared in only 56 games and was sent to the plate only 76 times (.230, five doubles, two triples), it was his heroic, two-out, bottom-of-the-ninth, pinch-hit single that scored Ott to win Game Two in the Series.

"God gave me that. He blessed me," Sanguillen insisted. "God knew what it meant to me to win, and to win as a Pirate, so I think He gave me that last moment to be a part of it. We were the 'Fam-a-lee' like everyone heard about and that's why we won."

Sanguillen also had the opportunity to witness Clemente's legacy live through Hall of Famer Willie Stargell after his return from the West Coast and to the "City of Champions." Not only had Stargell become known as "Pops" later in his legendary career, but, according to Sanguillen, the first baseman often echoed the lessons taught by Clemente.

"Winning the '79 World Series also meant a lot to me because without Clemente, Willie picked it up to show us how to win. It was beautiful. He was there all the way. It's amazing how the players listened and respected him no matter what. To this day, Clemente was the foundation of the Pittsburgh Pirates, and Willie took that and carried it for all of us. I thank God I was there."

For a third straight season, Sanguillen's time on the field declined in 1980, and although Pittsburgh finished 83-79 and in third place in the East Division, he was confident his diamond time was exhausted. In the years that followed his retirement, employment as a player agent and as a clinics coach preceded yet another return of the Sanguillen smile to Western Pennsylvania.

Sanguillen joined several former teammates in the late 1990s, including Dave Parker, Bill Virdon, Mazeroski and Stargell, at Pittsburgh's spring sessions in Bradenton, Florida, and beginning in 2002, the former All-Star became a fan-favorite fixture at PNC Park with the opening of Manny's Barbeque beyond the park's center field wall.

"And then I got my bobblehead," he said. "That was big for me because, in this world, you never know what you're going to get. My father always told me, 'You know where you are born, but you don't know where you're going to end up.' He was right.

"When I said I would come to the United States to play a game I knew nothing about, I never thought I was going to find people like Roberto Clemente or Willie Stargell – the kind of people who really bring out love and kindness in everyone they touch."

8

Dave Giusti

Pirates Reliever, 1970-76

"There was no one more disappointed with the loss than David Giusti, but David did exactly what Roberto told him to do – he picked himself back up and didn't quit."

— Steve Blass

Dave Giusti's search for an advantage against opposing batters led the right-handed hurler down a path not often traveled by a baseball player in any era of the sport. More – more power, more speed, more velocity – has long been preferred by the game's scouts and executives while scouring the globe for the "next best," but in Giusti's case, a change of pace proved unusually successful.

"The scouts were always looking for the flame throwers," Giusti said. "But, when you realize that you don't throw as hard as other guys, you have to figure out how to get outs. That's what I did.

"Some guys were throwing knuckle balls and some, like ElRoy Face, threw a split finger," he continued. "Well, the 'palm' ball was a pitch that didn't have as much spin as a split-finger, but it had more than a knuckle so you couldn't really see it coming."

Historically, the reliever is most known for his "palm" ball, a pitch he learned in the early 1960s while toiling for Syracuse University. The "palm" was Giusti's changeup in theory, but because of the ball-engulfing grip he employed, each offering would also dip away from the strike zone just before reaching the catcher. As a young professional with the Houston Colt .45's in 1961, he perfected the pitch. "I first started throwing it consistently for strikes that first season," he explained. "Then, after I came back after surgery (on the right elbow to remove a bone spur and bone chips) in 1964, it helped me get back up to the big leagues to stay (by 1965)."

"When he had it working good, it was unhittable," former catcher Manny Sanguillen recalled. "The arm speed was the same as his fastball so you couldn't tell it was coming. Then it died at the end."

Giusti was utilized as a starter during his first professional decade, making 125 starts for the Colt .45's/Astros (1965-68) and Cardinals (1969) before being acquired by the Pirates in October 1969 in a four-player deal. Once in Pittsburgh, he was transformed into a reliever, and a successful one at that. In 1970, Giusti collected 26 saves in 66 appearances, and he also served a vital role in the Pirates' championship season in '71. Giusti earned the National League's Fireman of the Year Award with a league-leading 30 saves and a 2.93 earned-run average.

"He had (the batters) off balance," Sanguillen explained. "They would think the fastball was coming, but we would throw the 'palm,' and they'd either pound it into the ground or swing and miss. It was a good, good pitch. Even when the batter knew it was coming, he couldn't hit it."

Until, however, Johnny Bench did. Pittsburgh appeared on its way to a second straight World Series

Dave Giusti – 85

when the Pirates possessed a 3-2 lead entering the bottom of the ninth of Game Five of the NL Championship Series. Manager Bill Virdon used reliever Ramon Hernandez to get the last two outs in the eighth following yet another masterful game by starter Steve Blass. Virdon then called on his closer to face Bench, first baseman Tony Perez and infielder Denis Menke.

"It was the same thing we did all year," Virdon explained. "Giusti was our guy in the last inning so it was natural for us to turn to him in that situation again."

In fact, it was Giusti who earned the save in Game Three after entering the 3-2 contest in the eighth. "I really never paid too much attention to the score when I went in because we usually had the lead and my goal was always to shut them down completely, not allow a run or two just because we had that size of a lead," Giusti explained. "So, when (Virdon) called on me in that game, I used the same attitude I always did. I went to the mound looking for outs."

Giusti worked Bench to a one-ball, two-strike count. "On the third pitch, Bench ripped a fastball down the left field line," the hurler remembered. "So, I threw him what I thought was a heck of a 'palm' ball."

The Reds catcher, however, smacked the offering over Riverfront Stadium's right field wall to knot the score, 3-3. "It was almost like he waited a little bit longer than what he used to, and he hit it hard. I was just absolutely dumbfounded. I just couldn't believe it. I had been so successful in that kind of environment where I knew he was looking for a fastball because I just threw him one.

"I just lost it – totally lost it. The next two batters? I don't know what happened other than I lost my composure."

Giusti surrendered singles to Perez and Menke before Virdon replaced his closer with starter Bob Moose, a right-hander who had made 30 starts and one relief appearance in '72. Three batters later, Moose tossed a wild pitch past Sanguillen that permitted the winning tally to reach the plate.

"All these bad things go through your mind," Giusti said. "I knew I let the team down, and I let the town down. That's the most devastating thing that ever happened to me in the game of baseball."

Mourning in the visitor's clubhouse, Giusti was confronted by the ballclub's leader, Roberto Clemente. The right fielder approached the pitcher and said, "Giusti! Damn you, Giusti! Look straight ahead. Pick up your head. Don't quit now."

"People can believe whatever they want about Roberto Clemente, but he wasn't just one of the greatest players I'd ever seen, but also one the greatest teammates," Giusti said. "He also grabbed me by my shoulder and said, 'Just remember, the most important thing is not this game. It is you and your family.' What he said to me after that loss got me back to believing in myself again.

"That loss – that one pitch – still haunts me, and I know that it's natural. And I know there are a lot of people who are going to remember the bad things instead of the good things. Look at Bill Buckner – all you have to do is look at his stats and you can see that he had a hell of a career, but all people are going to remember about him is that he missed that one ground ball (in the 1986 World Series). It's not fair, but what are you going to do about it? Nothing."

Blass also attempted to raise the reliever's spirits once the club returned to Pittsburgh. The pair traveled to their South Hills homes together with their wives, but during the trip Blass suddenly pulled the vehicle over and ordered a "Chinese Fire Drill."

"Steve was just trying to make me feel good," Giusti said. "We screamed and yelled and then we got back in the car and went home. It made me feel like people don't care as much about the game as they do about the friendship."

"Losing," Blass said, "is tough no matter when it happens. For us to lose that game that way? One of the best things about baseball is that there's always tomorrow – and another game – to help you forget about the losses, but this time there was no tomorrow. So we did what we had to do to get past it and move on.

"There was no one more disappointed with the loss than David Giusti, but David did exactly what Roberto told him to do – he picked himself back up and didn't quit."

The loss, though, prevented the '72 Pirates from proving their superiority over the title team that swiped baseball's championship from the Baltimore Orioles the previous autumn, and from establishing a dynasty similar to what was achieved in Oakland and Cincinnati in the 1970s. In a classic diamond duel, Pittsburgh scored a 2-1 victory over the Orioles in Game Seven following a four-hit gem by Blass. Giusti did not allow a run in three Fall Classic appearances, and recorded the save in the fourth game.

"We won the division in 1970, so there was a lot of expectations for the '71 team," Giusti said. "We had all these great players, and they were a year older and the ballclub was expected to be better – and we were, and we proved that by beating the Giants in the playoffs, and then beating the Orioles in the Series. Boy, that was a hell of a Series, and still, that was the most enjoyable season I ever had playing baseball.

"In 1972, the same rules applied. Clemente was back, and so were (Willie) Stargell and Blass and (Nellie) Briles and (Al) Oliver. The fans expected more from us, and we expected more. There was a proud factor involved when you were a part of the Pittsburgh sports scene then. The fans had the Steelers, who were starting to win, and they had the Pirates. It was an unbelievable time in Pittsburgh."

The Pirates finished the schedule 37 games above .500 in 1972, but the ballclub owned a 4-8 record against the Reds heading into the playoffs. "I think you can say that the '72 club was better than the team we had in '71," Virdon said. "But I think it was because of the experiences we had in '71. You can't go through the playoffs and a seven-game World Series and not learn something about being there. I think if we would have gotten back (into the World Series), that experience would have benefited us. To what extent? Well, that's impossible to say today."

Although Giusti received a baseball scholarship to pitch for Syracuse University, he also played third base and was a member of the Orangemen basketball squad from 1958-61. But after leading Syracuse to the College World Series he opted to sign with Houston for a $30,000 bonus that was paid over a three-season period.

"The Cubs wanted to look at me as a third baseman and the Cardinals saw me as a pitcher," he said. "But the reason why I had teams looking at me was not because of where I played – Syracuse University – because we only played 18 games the last year I was there. It was because we went to the College World Series my senior year."

Giusti did not stick in the big leagues until after fully recovering from the elbow surgery that caused him to miss the entire '63 campaign. In 1965, he made 13 starts and 25 relief appearances for the Astros, and he led the club in wins with 15 the next year. Although his ERA was 4.20 in 210 frames and 33 starts in 1966, Giusti hurled a one-hit, 1-0 win over the Giants and blanked the Reds, 11-0, in August alone.

The Astros, however, did not post a winning season

until after Houston traded Giusti to the Cardinals prior to the '69 season. "We had a decent club, but not a great club," he said. "We had a lot of young kids who you could see that were coming up – Larry Dierker, Joe Morgan – but we didn't hit much.

"When I got traded to the Cardinals, I knew it was an opportunity to be on a first-class ballclub, and I was happy to get there. But then I came down with a real bad back and was on the disabled list for about 45 days. I didn't know how to handle that because I was never on any disabled list. That hurt me quit a bit with the Cardinals."

His 3-7 mound mark and 3.61 ERA were not enough to stop St. Louis GM Vaughn P. "Bing" Devine from dealing the righty to Pittsburgh along with catcher Dave Ricketts for utility player Carl Taylor and minor-leaguer Frank Vanzin. Pirates' general manager Joe L. Brown was searching for a starter to complement Blass, Moose and Doc Ellis. Frequently, Brown consulted his All-Star right fielder before pulling the trigger. "From what I understand now, Clemente said, 'Get that little chubby Italiano on our club,'" Giusti explained. "I've been told that Roberto didn't like facing me for whatever reasons, but I guess Joe listened to him.

"But, when I got to Pittsburgh, I was made a reliever because I had a horrible spring training. I wasn't upset with the decision because I figured if I concentrated on getting people out, good things would happen. I believed that if I got the job done that I could make a career out of it."

Giusti admitted to "pressing" to impress too much following his disappointing season with the Cardinals. "It's impossible to know why. There are a lot of variables that come into play. Sometimes it's because you have good stuff and you're getting hit. You make a good pitch and you're getting hit, so you start to doubt your abilities. That was biggest thing I had. When I was throwing good and still getting hit I questioned my abilities in the game. It's a common thought process that goes on with everybody in every job at one time or another.

"But then I found myself in situations to save games, and I got the job done one night, and then on another night, and then another," Giusti explained. "Then I quit pressing and started pitching."

Giusti, an active member of the Pirates' Alumni Association since his retirement, joined the Pittsburgh organization at 30 years old, and soon made friends with the 28-year-old Blass, who, in 1970, was waging his own war for major-league establishment. The pitchers were soon neighbors and road-trip roommates, and the pair also celebrated victory and mourned defeat together. The 1973 season, however, offered more than either could handle.

Not only was Clemente tragically killed in a plane crash on New Year's Eve 1972, but the outstanding control Blass was known for also vanished. The pitcher finished second in the Senior Circuit's 1972 Cy Young voting after posting a 19-8 record and a 2.85 ERA, but he proved to be a liability in 1973 because of an inability to throw strikes. By spring 1975, Blass was forced to call it quits.

"I tried my damnedest to try to find out what was going on with Steve, but it was hard for me to intercede too much because I didn't want to lose him as a friend," Giusti explained. "He was so solitary with the way he handled it. We were roommates and he would leave at 7 a.m. sometimes. He would be going to see someone, but he wouldn't tell me. It was hard for me to handle that, but I knew all I could do was be his friend and be there for the times when he did want to talk to me about it."

Blass spent the majority of his 1974 season with the

Charleston (West Virginia) Charlies, and although he failed to improve his pitch location, Brown and skipper Danny Murtaugh allotted the right-hander one last outing during the team's next spring session in Bradenton. "It was sad," Giusti said. "I think it's one of the saddest days I ever experienced in the game.

"To this day, I don't know how he did it, but afterward Steve made a statement about the game of baseball and that he just couldn't pitch anymore. There were a lot of guys in tears. I'll never forget that. He never had a sore arm and he added so much to the club on the field and in the clubhouse, and he had to go for reasons we've never found out about."

"Dave Giusti was there for me," Blass explained. "I can't explain it although I lived it, but I can tell you that Dave was there every time I needed him to be. He tried to tell me the same thing Roberto told him after that loss (in the 1972 NLCS) – to stand up and not quit."

Blass' inconsistencies and the loss of Clemente derailed the Pirates in '73, and although Giusti registered 20 saves, a 2.37 earned-run average, and was named to the NL's All-Star team, the club was 80-82. "I think we all still had visions of Clemente in right field, and really, there are no reasons why we didn't win our division. We lost it," Giusti said. "Probably the ghost of Clemente had something to do to it. We were a much better ballclub than our record."

Pittsburgh rebounded in 1974 and '75 to win the East Division, and Giusti went 12-9 with 29 saves in 125 appearances, but he and the club lost to the Dodgers and Reds in the NLCS, respectively. After his six-save, injury-plagued season in 1976, the 36-year-old Giusti was traded in an eight-player deal that brought Phil Garner to the Pirates' roster. In August 1977, the A's dealt the hurler to the Cubs. "I was devastated. I could have appealed the trade (from Pittsburgh) because I had 10 years in the majors and five years with the same club, but I didn't. I thought, maybe, someone else wanted me – but I could see the end was near.

"Later that season, the A's dealt me to the Cubs," he explained. "I was ready to quit then, but then Steve (Blass) came to Chicago. We went out and had a blast. That allowed me to stick it out until the end of the season. When the last game was over, I knew that my career was, too."

Giusti, employed by Marriott Corp., and then American Express before he retired in 1995, explained he took the positives instead of the negatives of a 15-year career with him when taking his walk away from baseball. "There are a lot of single things you remember, but how could you top winning the World Series with the best team in the world?

"No matter what happened through the years, I made a decision a long time ago to take the good memories and to forget about most of the bad. There's a few that's going to get under your skin – like that pitch to Bench – but winning is so good it's able to erase a lot of it. Winning is that good."

9

Nelson Briles

Pirates Pitcher, 1971-73

#34

"Nellie went out there and pitched to win, that's for sure. But no matter what happened, you'd hear him singing something in the clubhouse after the game. Most days, you knew he was doing what he was meant to do for a living because he was usually better than most in the game, but every once in a while, you'd wonder if he should have been a singer – he had a pretty good voice."

— Bill Mazeroski

Nelson Kelly Briles, a California kid who held a wild wish to become a big-league starting pitcher, owed Roberto Clemente much gratitude before ever meeting the Pirates' Hall of Famer. It was Clemente, after all, who fractured St. Louis hurler Bob Gibson's right fibula on July 15, 1967, thrusting Briles into the Cardinals' rotation.

Briles, a part-time starter who was used by St. Louis manager Red Schoendienst more often out of the Redbirds' bullpen since debuting in the majors as a 21-year-old right-hander, immediately replaced Gibson.

After losing his first start, Briles reeled off nine consecutive victories and completed the campaign with the National League's highest winning percentage (.737, 14-5), the club's lowest earned-run average at 2.43 – and with a World Series championship ring on his finger.

The '68 season saw Briles make 33 starts and register a 19-11 mound mark as the Cardinals returned to the Fall Classic only to lose in seven games to the Detroit Tigers. During the next two seasons Briles was 21-20, but in 1970 the control artist's ERA inflated to 6.24 in 19 starts and 30 total appearances.

He became expendable as St. Louis general manager Vaughn P. "Bing" Devine prepared for the mini-dynasty's fall and the rise of Cincinnati's "Big Red Machine," and it was Pittsburgh's Joe L. Brown who expressed interest in acquiring Briles after a conversation with his All-Star right fielder.

"I found out 20 years later who the man most responsible for me coming to Pittsburgh was – Roberto Clemente," explained Briles, who had signed with the Cardinals in 1963. "Joe Brown went to Roberto and told him he was looking for a pitcher. He told him he wanted a veteran pitcher with World Series experience.

"I was told Roberto didn't hesitate when saying, 'Nellie Briles.' Joe asked why and Roberto said, 'because he wasn't afraid to knock me on my ass. He's not afraid. He competes. He's the type of guy I want.'"

The Pirates captured the NL Eastern Division title in 1970, their first championship since upsetting the Yankees for the 1960 world title, but the Reds swept Pittsburgh in the NLCS. Brown then set out on a search for a hurler to replace southpaw Bob Veale in Pittsburgh's rotation and believed Briles could be the proper replacement. "Nellie was tough and he threw a lot of strikes," said starter Steve Blass. "We had a very good season in '70, but we were short. We had some

Nelson Briles – 97

great players on the club, but there were holes to fill."

Briles, told by Brown of his conversation with Clemente 20 years after the transaction was complete, recalled his strategy against one of the greatest players to don a big-league uniform. "I pitched him hard and inside. I really had to go inside to challenge Roberto. He didn't have many holes in his swing and I wasn't an overpowering guy," he said. "I relied on getting the ball in play. I was very aggressive inside, and I hit him a couple of times and there were times when I did knock him down.

"If you put the ball all over the plate against him, Roberto would nail you – he'd try to hit it up the middle on purpose like he did against Gibson so he could hurt you. He wanted to win, first and foremost, and if he knew you played the same way, he respected you."

Once Brown's offseason dealings were complete, the 1971 team featured an offense with veterans Clemente, Manny Sanguillen, Willie Stargell and Al Oliver, and youngsters Frank Taveras, Rennie Stennett, Milt May and Richie Zisk. The pitching staff included hurlers such as Blass, Briles, Bob Moose, Dock Ellis, Bruce Kison, Bob Johnson and Luke Walker.

"What was unique about that ballclub was, when you look at all the key players, they were all organizational players and they were all young players," Briles explained. "Most of the club's players were all home-grown, but there were also veterans like myself that gave the team experience and stability."

When Pittsburgh opened the schedule at home against the Phillies, however, Briles was not among manager Danny Murtaugh's starters. A multi-player deal with Kansas City brought the right-handed Johnson to the Pittsburgh roster and he, instead of Briles, was implemented into the rotation. Briles was remanded to the Bucs' bullpen. "I didn't know at the time of the trade that I wasn't considered one of the starters, and that didn't sit well with me because I wanted to come over and be a starter," Briles said. "I thought because of experience I would, and I was disappointed. After a while though, we were winning and I played an important role."

The Pirates recorded a 12-10 record in April but then caught fire in May, June and July to enter the season's final two months at 26 games above .500 and in command of the division race. The club, which scored the most runs in the NL with 788, rebounded from a 14-17 performance in August to finish 16-9 in September and seven games ahead of St. Louis for the Eastern Division trophy.

Briles made 37 appearances – including 14 starts – and owned a regular season record of 8-4 with a 3.04 ERA, but his final outing of the schedule jeopardized his mound time in the postseason. "I spot started most of the season and ended up being the go-to guy in September and pitched very well down the stretch in real key ball games," he said. "At the end of the season, Murtaugh told me that he was going to count on me in the playoffs and that he wanted me to get some innings in against Philadelphia during the final series of the season.

"I told him that I didn't need it and that I would throw on the side instead because he said I would probably get a start in the championship series against San Francisco. Well, I ended up starting against the Phillies, and I ended up straining my groin."

Murtaugh scheduled Briles to begin the third game of the championship series against the Giants, one of only two teams the Pirates registered a losing record against during the course of the '71 campaign. However, when game time arrived, the right-hander was met with a moment of personal definition. "We had not played San

Francisco very well – we were 3-9 against them in 12 games," the hurler recalled. "The series was tied, 1-1, and we were back in Pittsburgh for Game Three, and I felt fine during the warm-up in the bullpen – until I tried to give it the extra 10 percent. That's when I felt the grab.

"I stepped off the rubber and knew I had to make a decision. I remember thinking to myself, 'You can't be selfish. This ballclub has a chance.' That's when I called Danny and told him I was 90 percent. I told him I was more than willing to go out there and that I could at 90 percent. He decided to call on Johnson, and he ended up pitching the game of his life.

"That was one of the toughest decisions I ever had to make in my baseball career because I wanted to pitch. That was my chance to vindicate – in my mind – for being overlooked early in the season. That was in the back of my mind, it really was, but I knew what was the right thing to do and really didn't hesitate to put someone else in who was 100 percent. It was absolutely the right decision, and we ended up winning the series and facing the Orioles in the World Series."

Baltimore fashioned a 101-57 record in 1971 and approached the World Series fresh off a sound sweeping of the Oakland A's in the American League Championship Series. The Orioles were powered by an offense highlighted by five regulars with 20 or more home runs and four starting pitchers who won no less than 20 games and lost no more than nine. "They were the best in the American League," catcher Manny Sanguillen said. "And we were the best of the National League. It was going to be a fight, and no one was backing down."

Ellis lost to left-hander Dave McNally, 5-3, in Game One, and Johnson was creamed, 11-1, by Hall of Famer Jim Palmer in Game Two. Blass, who materialized as one of the Pirates' Series heroes, tossed the first of a pair of complete-game victories in the third contest, and Walker, Kison and reliever Dave Giusti combined for a four-hit, 4-3 decision in Game Four to knot the classic battle at two wins apiece.

"Murtaugh told me I would start the fifth game, which pleased me very much," Briles recalled. "And the good Lord smiled on me and I pitched one of the best games of my life."

In nine frames in front of 51,377 fans at Three Rivers Stadium, only two Orioles reached first base via two singles. Not one Baltimore bird reached second as Pittsburgh – and Briles – claimed a 4-0 triumph for a 3-2 lead. "I didn't feel the vindication I thought I would," he said. "But I did believe at the time that I re-established myself not only as a competitor but as a guy who could rise to the occasion – a guy who, when all the chips are on the line, you could count on.

"I wouldn't say there were ill feelings at the time, but I didn't think I got a fair shot out of the shoot, and that's why winning that game was self-gratifying. Winning for the ballclub and putting us ahead in the Series in a pivotal game turned out to be pretty important for us. To pitch that big of a game, and one of the biggest games in World Series history, was very special."

Briles had quickly met greatness face-to-face after spending just one season in the Cardinals' minor-league system in 1964. In his six years with St. Louis, not only was Gibson one of his mound mates but the Redbirds' roster also consisted of Hall of Famers Steve Carlton, Lou Brock, Tim McCarver and Orlando Cepeda. Following his departure from Pittsburgh after the 1973 season, he also counted amongst his teammates George Brett, Gaylord Perry and Palmer.

But no one was greater than "The Great One."

By the time Briles landed on the Pirates' roster, Clemente had appeared in 10 All-Star games, and collected 10 Gold Gloves, four Senior Circuit batting titles and was named the National League's most valuable player in 1966. Despite the awarded hardware and decorations, Briles still possessed an opinion formed only on heresay. "I only knew what I knew from playing against him for six years, so I thought what a lot of people thought – he was a hypochondriac and he was controversial.

"I heard he didn't want to play hurt, and that's one of the reasons I didn't have trouble getting up to play against him. I used that stuff for motivation," Briles continued. "Then I met him and I watched him play every day and I came to realize that all the demons that he chased through his early career – the lack of recognition, the lack of respect, the difficulty of being not only black but of being Latin American – were all behind him. He had won all of those battles."

Briles' first man-to-man meeting with Clemente occurred on the hurler's first day in Pirates camp in Bradenton, Florida, when he approached the outfielder in the McKechnie Field clubhouse. "I had thrown at him and battled with him at the plate in the past, but I knew I needed him on my side. So I just went and offered a handshake. He was over at his locker with his shirt off – I tapped him on the shoulder and he kind of turned his head a little.

"I said, 'Roberto', and I stuck out my hand, 'Nellie Briles.' I didn't know he had a sense of humor, but he just looked at me without a smile on his face and said, 'I know who you are.' I said, 'Roberto, all I ask is you catch a couple of my mistakes this year.' He said, 'Nelson, you forget I watched you in St. Louis. If you want me to catch a few of your mistakes, you better learn to keep them in the ballpark.' He started laughing like hell, and that was it from that moment on.

"Then I watched him play, and I knew everything I had

heard about him was nothing but bull. He did things on the field that no one had done before, and no one has done since."

Clemente's death on New Year's Eve 1972 shocked and saddened Briles to the same extent the tragic plane crash did to all the right fielder's teammates, family and friends, but he was not surprised when informed the reasons why the right fielder felt the need to accompany the medical and relief supplies from Puerto Rico to the earthquake-ravaged Nicaragua. "Roberto knew the supplies would just end up on the black market and never get to the people just like the other supplies he had sent before that flight," he said. "He wanted to make sure it got to the people. That was the dedication. He wanted to make a difference with his life outside of baseball."

The following spring training "did not feel right," Briles explained, and neither did Opening Day nor the entire 1973 season. Sanguillen, Clemente's closest friend on the club, moved from behind the plate to replace No. 21 in right field at the request of Roberto's family, but no alterations could serve as proper replacement. "When that season opened and you didn't see No. 21 run out to right field, the reality of it all really set in," Briles said. "There was no replacing Roberto Clemente, no matter what. He was the leading force on the ballclub, and we floundered without him."

Briles, though, was 14-13 with seven complete games and a 3.26 ERA, and although the club finished 80-82 and in third place, he expected to be brought back the next season. From 1970-73, he compiled a 36-18 record with a 2.98 earned-run average in 74 starts and 98 appearances.

Instead, Brown swapped Briles to Kansas City along with infielder Fernando Gonzalez for utility player Ed Kirkpatrick, infielder Kurt Bevacqua and farmhand Winston Cole. "When I was traded to Kansas City, I couldn't believe it because I thought I was a pretty good citizen in this town and represented the club real well," he said. "They told me at the time that they were looking for hitting depth, but all they got was another back-up catcher and utility-type guy. I didn't want to go, but I had no choice."

Two years with the Royals ended with a trade to Texas, and after a full season as a starter, Briles again returned to split duty between the bullpen and the Rangers' rotation in 1977. In September that year, Baltimore bought the right-hander, but after a nagging Achilles' tendon strain during the '78 campaign limited his innings, Briles called it quits. "I did go to spring training with the Mets the next year, but I decided at the end of spring that it was time to get on with the rest of my life.

"I didn't know exactly what that would mean for me at the time, but I knew my playing days were over," he said. "I was only 34 when I was done playing, but I also got up to the majors pretty quick and pretty young, and I was out there for a while. It took its toll."

Upon his return to the Western Pennsylvania region, Briles investigated several private business opportunities, but he also called decision makers at KDKA-TV and was soon employed as a broadcaster during Pirates games during the 1979-80 seasons. "I don't think that surprised anyone who was ever on a team with Nellie," Hall of Famer Bill Mazeroski said. "As a player, Nellie went out there and pitched to win, that's for sure. But no matter what happened, you'd hear him singing something in the clubhouse after the game. Most days, you knew he was doing what he was meant to do for a living because he was usually better than most in the game; but every once in a while, you'd wonder if he should have been a singer – he had a pretty good voice."

"The man sang the National Anthem before a World Series game in 1973, and he had a good voice," Giusti said. "He was a smart man, and he was talented – off the field. He was a good broadcaster, and not just because he played on the major-league level."

Briles then left the Pirates' TV team for USA Cable Network's Game of the Week for the 1981-83 campaigns and eventually ended his broadcasting career with a three-season stint as a Seattle Mariners announcer.

"The best thing that could have happened," Giusti said, "was that he finished there, came back to the Pittsburgh area and started working for the Pirates front office. Nellie Briles is the one and only reason the Pittsburgh franchise has the strongest alumni association in all of baseball."

He was initially offered the directorship of corporate affairs in 1986, and soon after he founded the franchise's Alumni Association – a non-profit organization that donated more than $1.3 million to Pittsburgh-area charities in its first 16 years of operation.

In the early 1990s, Briles initiated the Pirates' fantasy camp program and was named vice president of corporate projects in 1999. "Nellie was traded by the time I got up with the Pirates," former closer Kent Tekulve said. "But I've had the chance to work with him as far as the alumni association and the fantasy camps are concerned, and both are first-class operations. You hear about former players taking positions with their old clubs, but most of the time they are token positions and the guy doesn't have to do much to earn his money because of what he did as a player. That's not true with Nellie."

"When you meet him these days," Mazeroski said, "you'd never know he used to be a player because of the way he carries himself as a professional, corporate-type guy. You'd never know he was one hell of a tough pitcher in his day, but what he's done after his playing career is more than most of us could ever dream of doing."

The organization recognized Briles during the 2003 season, awarding the former pitcher with the only honor it bestows on an annual basis. On April 8, 2003, Briles was presented the "Pride of Pirates Award" by Kevin McClatchy, the club's CEO and managing general partner.

"Playing for the Pirates was a pleasure, and so has everything that I've done since coming back to the organization in 1986," said Briles, who is also a member of the Pennsylvania Hall of Fame's Western Chapter and Pennsylvania Sports Hall of Fame. "But doing what I've done since coming back to the organization has really allowed me to come full circle with what I wanted to do as a person, too. The awards have been great, especially receiving the Pirates' award. It's let me know that I've been able to do more than throw a baseball for strikes for a major-league organization."

10
Kent Tekulve
Pirates Reliever, 1974-85

> *"Tekulve took more to the mound than I ever did, and his submarine motion was part of it. There's a reason why we haven't seen a lot of guys use that kind of motion, and it's because there aren't too many who have figured out how to make it work for them. Tekulve sure did, though, didn't he?"*
>
> — ElRoy Face

Kent Tekulve lived long as the odd man on a baseball field because of a peculiar pitching motion and a slim physical stature. At 6 feet 4 inches and 155 pounds, the lanky Marietta College graduate was a sidearmer with little big-league potential when attending a Pirates try-out camp in July 1969 at Forbes Field.

"From what I've been told, he didn't look like much back then," said Chuck Tanner, one of 10 former managers who came to count on Tekulve during the right-hander's 16-season major-league career. "The scouts liked the big, bulky farm boys who could throw the ball 100 miles per hour, and Tekulve was tall and thin, and he didn't throw the ball very hard. That's why there weren't a lot of people who gave him much of a chance to make it out of the minors, let alone to do what he did with the Pirates.

"By the time I became the manager in Pittsburgh, he was in his third season in the major leagues, but there still wasn't much to him. And he wasn't like a lot of the other guys we had on the teams back then because he was quiet, he kept to himself and most people left him alone."

The drug scandal that flooded Pittsburgh and drowned Pirates fans in the early-to-mid 1980s served as a perfect example of Tekulve's solitary existence beyond the Astroturf of Three Rivers Stadium. In exchange for immunity, three former Pirates – John Milner, Dale Berra and Dave Parker – joined Tanner and fellow big leaguers Lonnie Smith, Keith Hernandez and Enos Cabell as witnesses during the federal trial of Curtis Strong. Several more Pittsburgh ballplayers were implicated during the 12 days of testimony, including Willie Stargell, Bill Madlock, Lee Lacy and Rod Scurry.

Parker, Berra and Milner all admitted to purchasing cocaine from Strong between 1980-83 – Milner even informed the court that one sale had taken place in the restroom area of the Pirates' clubhouse during a game – and Strong was convicted of 11 counts of cocaine distribution on Sept. 20, 1985. Federal Judge Gustave Diamond sentenced Strong to the maximum prison term of 165 years and fined him $275,000.

"And I guess I was in the middle of it and didn't even know it," Tekulve said with a laugh. "I was traded by Pittsburgh before the trial and all the publicity the case was getting in Pittsburgh, but I was there when this guy was selling it inside our clubhouse? I didn't know a thing about it.

"Maybe I was the most naïve person in the world because I didn't know the majority of this stuff was going on. I had been playing major league baseball for a long time, and never once did I have someone come up to

Kent Tekulve – 109

me and try to sell me cocaine. Maybe I was a bad target – who knows? – but I remember thinking that there must have been something wrong with me because I never got the offer. Maybe I was too big a square back in those days, I don't know, but I wish they would have asked so I would have been able to tell them no.

"A great example of how in the dark I was about all of this at the time was my relationship with Rod Scurry. There's a guy who ended up dying of a cocaine problem (in 1992 at the age of 36), and I had spent a few years sitting right next to him in the bullpen, in the clubhouse and on all those planes – and I had no idea," Tekulve continued. "I didn't know he had a problem. That scared me a good bit because I was in the middle of having kids at the time, and if I couldn't tell Rod had that big of a problem, how would I ever know if my kids did?"

Former general manager Harding Peterson traded Tekulve to Philadelphia for Al Holland soon after the 1985 regular season opened in order to avoid a public-relations backlash following the impending sale of the Pittsburgh franchise by the Galbreath family to a civic-laden ownership group. The new owners planned to gut the Pirates' payroll of all high-priced veterans, and although most of the heroes that captured the 1979 championship had either retired or since moved on via free agency, Tekulve remained on the roster.

"My first thought after I was told of the trade was that something was up because no one ever gets traded after the first month of the season," recalled Tekulve, a father of four and husband to Linda. "What happened, I learned later, was they were trying to work out a deal with Philadelphia during spring training and couldn't get it done. A bunch of factors were inter-related, although I didn't know about any of them at the time.

"Peterson knew about the sale of the team, though, and he knew what the new owners were going to do. The new guys wanted the Galbreaths to trade me first before they took over so they wouldn't become known as the bad guys in town," the former reliever explained. "Plus, I was getting close to becoming a 10-year veteran with five years with the same team, and that means I could have vetoed the trade. It was 28 days before that point in my career – and I'm sure Peterson knew that, too. All those factors made me the odd man out."

Tekulve immediately disapproved of the transaction, but with tied hands he joined the Philadelphia organization in a move that likely extended his career. "I was a 'Bucco' for life then, and I still feel the same way today," he said. "I didn't want to go anywhere, but when I look back on it, that trade probably put three or fours years onto my career. I would have been in the fire sale that took place after the new owners took over, and at that point in time who knows what would have happened. I went to Philadelphia and finished out my contract, got the extension and got another one-year contract out of it. For me, it turned out to be good."

Following his three full seasons with the Phillies, Tekulve was released and free to sign with Cincinnati, his true hometown team that had long dismantled the dominant "Big Red Machine." His tenure was short – 37 appearances for 52 innings in 1989 – and at the All-Star Break, Tekulve officially retired at 42 years old.

"When I signed with Cincinnati, (John) Franco was closing and (Rob) Dibble and Norm Charlton were kids and they needed someone to come in and be a buffer for them until they got their feet on the ground. I was the guy in the middle. But by the middle of the season I realized that the team was going nowhere and I, and my family, would be better off if I finally came home and stayed there for a while."

Despite his awkward motion and slender build, Tekulve was also often the last man standing, a trend initiated with that first tryout camp that continued throughout his entire career. But if it were not for longtime Ohio Valley baseball scout Dick Corey, the sidearmer may have entered the American workforce honing a much different skill than the slider.

Tekulve was invited by Corey to take part in the invitee-only camp after the right-hander was not drafted following his senior season at Marietta. In the morning, those operating the tryouts instructed Tekulve to throw off a mound, and then to run a 60-yard dash. He was soon dismissed.

"By the time Dick showed up, I was already sitting in the stands watching," Tekulve said. "He then shows up and finds out that I wasn't one of the guys they picked to keep around for the afternoon game. After talking to the people running this camp, he finds out that my time in the 60-yard dash wasn't good enough.

"After the afternoon game was over, there was another game scheduled with the 'Little Pirates,' a youth team the Pirates sponsored. Well, I ended up throwing to their catcher."

Peterson, who served as the organization's farm director at the time of Tekulve's audition, received a message that the team's Low-A affiliate in Geneva, New York, had lost a pair of pitchers to injuries. "I learned later that he didn't have anyone in Bradenton who was ready to move up to the New York-Penn League at the time, so he signed me. I'm assuming now that the thought process was that I was older, out of college, threw a slider and maybe I could get by and help them get through the season," Tekulve continued. "That's when they signed me. If I had just walked out of the tryout camp and went home like they told me to none of this would have ever happened. But there I was, the last guy standing."

As Tanner noted, Tekulve may have signed a professional contract, but expectations for his future were low. His pitches lacked the preferred velocity, often topping out in the low-80 mph range, and most of the game's top hurlers at the time owned textbook, over-the-top motions. Instead of being referred to as a "prospect" by team officials, Tekulve recalled, the right-hander was used in a variety of roles as long as his presence did not inhibit the progress of blue-chip players. "That's how I became a reliever in the first place," he said. "I was first a starter and put up some pretty good numbers, but then they made me a reliever because they didn't want to delay any of the prospects. They figured that I'd go up there and screw up and then they could get rid of me.

"But I kept putting up decent numbers, made a couple of All-Star teams and kept moving up until I got to the majors. It's really amazing – I was within a hair of never playing one game of professional baseball. The funny part was from the day I signed all the way through the minor leagues, I wasn't playing because my sole purpose was to be a major-league baseball player. The reason I signed was because I was basically going to make the same amount of money ($450 per month) the rest of the summer, but I was also going to get to play ball seven days a week and I was going to get to play against better players. I never really thought of myself as a major-league player all the way through. I just thought it was great because I was single, I was out every summer in a different city playing ball against the best competition I could find, and I was getting paid for it."

Tekulve pitching patterned his style after former Cincinnati pitcher Ted Abernathy, a sidearmer who spent 14 years in the majors with the Senators, Indians, Braves, Reds, Cubs, Cardinals and Royals. Just as Abernathy, who paced the majors in appearances three times and in saves for two seasons, allowed his motion to change as necessary, Tekulve's delivery soon sunk below the zero-degree arm angle once he reached the Double-A level. "What was happening was the ball was in the strike zone – the legitimate strike – but the ball was on the outside of the zone and wasn't moving. The hitters were hitting it, and they weren't swinging at as many of the balls off the plate but hitting more of the balls in the strike zone.

"So I did what every pitcher does – I changed," Tekulve said. "I needed more movement so I tried different things. I tried pitching with a three-quarter motion, and that was nothing but a disaster. That didn't do anything. It was 81 mph and was dead flat. I then taught myself how to be a submariner because I had to figure out something. When I did it, I got the movement I wanted on my fastball. I just had to learn how to throw a curve and a slider, and once I did I was called up to the big leagues."

In 1974, the Charleston (West Virginia) Charlies served as Pittsburgh's Class AAA affiliate, and after Tekulve posted a 6-3 record in three starts and 33 relief appearances, he was summoned to Pittsburgh. In eight games and nine innings, he was 1-1 with a 6.00 ERA. Only then did the Pirates' all-time leader in games doubt his improbable dream. "It reminded me of what Peterson said to me after I first joined the Geneva club my first year. He came over to me and Bruce Kison and tried telling us sidearm pitchers didn't have a future in the big leagues because of the Astroturf that was coming into the game at the time, and because they were lowering the mound because of Bob Gibson.

"At the time, I decided to keeping doing it, but after that short promotion with the Pirates, I thought about it again. I hoped I made the right decision, and I took that hope to spring training with me the next year. After we won the World Series, one of my favorite stories to tell to the fans is the one about Peterson being wrong about both of us – we were both pitchers on that championship team, and we both had decent major-league careers. Just goes to show you how smart you have to be to be a general manager of a world championship ball team."

This "Rubber Band Man" soon became known as "Teke" among teammates and fans, and his crown-like Pirates cap was speckled like a sparkling galaxy with "Stargell Stars" as the closer took the Memorial Stadium mound in the eighth inning of Game Seven of the 1979 World Series with a 2-1 advantage. At the moment Omar Moreno caught the 27th out, Tekulve raised his arms in triumph with his third Series save and was immediately engulfed by every member of the club. Stargell was named the most valuable player, but Tanner insisted his submarine reliever never received his due credit.

"For a guy more suited to be a set-up guy, he didn't do too bad as a closer, did he?" Tanner said with a smile. "That postseason, I used him as the set-up guy for Don Robinson against the Reds, and as our closer in the Series against the Orioles. It depended on the lineups and how I thought we would get through the later innings without too much trouble, and that's all that mattered. 'Teke' didn't care when he pitched. He just wanted the ball whenever it was time for him to get the ball.

"He liked to pitch a lot, and his arm was great," the former Pirates' skipper continued. "In that 1979 season alone, that young man went out there 101 times – 94 in the regular season and seven more times in the postseason. He may be surprised I know all those statistics off the top of my head, but I do because I appreciated what he did for us. Everyone on that club appreciated him."

Tekulve had evolved into the Pirates' closer because he won a contest without contenders. He first appeared in the role in 1976 after Dave Giusti suffered a back injury and Bob Moose developed a blood clot in his right shoulder, but the acquisition of Rich 'Goose' Gossage in 1977 pushed the sidearmer over to set-up duty once again.

"In 1976, I became the closer by default. I was the last man standing again," he said. "In '77, Gossage is the closer, I'm the setup guy, we ended up setting the club record for appearances with 72 each, and I end up 10-1. To this day, I can't figure out how I lost one. I was the setup guy and I had Goose Gossage behind me? What more could you ask for?

"So I got to be known as a good set-up guy, but then Gossage goes to the Yankees. (The Pirates) acquired (Jim) Bibby to do the job, but after the first month of the season he wasn't getting the job done. So who do you think was the last man standing?"

Tekulve moved into the role and finished the '78 campaign with a club record 31 saves in 91 appearances as Pittsburgh finished second in the Eastern Division. He saved both ends of a doubleheader against Philadelphia on Sept. 29 that season, and was named the Senior Circuit's Player of the Month in August. While many moves were made in the following offseason, Pittsburgh's front office did not attempt to upgrade the club's late-game relief. "We made a lot of moves and made a lot of changes, but we knew we could count on 'Teke'," Tanner said. "If his stuff was working, and it usually was, he was one of the hardest to hit in the game.

"Even though his velocity did increase as he got older, 'Teke' never did fit the formula. He wasn't a fireballer like a lot of other closers were ... like Gossage was for us in '77, but outs were outs, and that's all we cared about. That's Pirates baseball."

Right-hander ElRoy Face was the Pirates' first closer

when he became a late-inning specialist in Pittsburgh during the late '50s and 1960s, and Giusti and Gossage followed. "The Pirates have always been one step ahead with their bullpens," Tekulve said. "Historically, we were always one step ahead, and ElRoy Face was one of the pioneers. Not until ElRoy showed the value of a true reliever did it become a job on a ballclub. Then teams had to have both the set-up guy and the closer because they became necessities."

Tanner's usage of Tekulve in '79 was unprecedented, but the reliever insisted he made so many appearances not because of guts, but because he never owned a reason to say no. "I never thought about it. What I did was I pitched every day and I just did it. I didn't do anything special to allow me to pitch every day. I didn't have some secret conditioning program. I was just able to pitch every day, and I was able to pitch two or three innings every day. It didn't make any sense but it worked. I just was always that way.

"Me and Chuck had a running joke about him asking me if I was good to pitch. One day I stuck my head into his office and told him that if I couldn't pitch I'd tell him. That way I knew he could use me any way he wanted – and we all know Chuck Tanner never managed by the book. He didn't go by the book because he never got a copy of it. He would do things on hunches sometimes. He would run at the wrong time just to have them not know when you are going to run, and he would pitch guys in different spots just to throw the other team off.

"That's how a lot of things worked out that '79 season. I don't know how many of Chuck's hunches worked out or didn't," Tekulve continued. "And most managers would pull their hair out if they had that many new guys, but Chuck didn't try to mold the team. He let the team mold itself. If it didn't work, he changed the beat. All I do know is that I wear a pretty nice ring, so that tells me Chuck Tanner knew exactly what he was doing."

Disco had yet to die in 1979, and Peterson, who was promoted to general manager prior to Opening Day, collected a ballclub that included players hailing from five different countries. Flamboyance was the trend, and in Pittsburgh, the beat was heard at Three Rivers Stadium. There were clubhouse fights and heated, chest-touching arguments, but there was also that rare, in-sync, on-field feeling. "We may have screamed and yelled at each other, but not during that three-hour period when we had a ballgame. We didn't feel like we needed to win the games, we felt that teams needed to beat us," Tekulve said.

"Sure, there were fights and things – and I was the guy who always got picked on the most because I was the skinny guy! The great part about it was that I knew, no matter how much harassment I got all day long, when it came time for the ninth inning, I was the guy they wanted on the mound. It was a unique ballclub. We used to call ourselves the 'United Nationality Baseball Club' because we had about one player from every nationality. Nobody cared where anyone was from, and that was because of Willie."

Stargell, the franchise's all-time power producer with 475 home runs and 1,540 plated runs, was the leader in the locker room, earning the nickname, "Pops," during the latter portion of his career. It is Stargell, according to Tanner, Tekulve, Sanguillen, Parker and every other member of the championship team, who rests still as the reason for the "FAM-A-LEE."

"Leaders put up good numbers – that's always the first rule – and Willie certainly did that for that team," Tekulve said. "But that man's influence on that club was huge. Because there were all the different personalities and nationalities in that one room, it could have gotten out of control. The potential was there, and there were those times when it could have all boiled over – but it didn't.

"It all clicked on the field, and that was Willie's doing. He kept that part of it together. You just watched the way he went about his business. After the games, you could look over and you couldn't tell if he was 0-for-4 and struck out four times or hit four home runs. He was on such an even keel all the time – when this one's over, you do what you got to do, talk to the media, take a shower, turn the page and get ready for the next one, whether it was good, bad or otherwise. That was the influence – that steady, consistent influence was the big part of Willie and his leadership."

It was also Stargell who adopted the identity of the 1979 Pirates during a mid-season rain delay in Pittsburgh. Several members of the club remained in the dugout as the Three Rivers grounds crew covered the park's artificial surface, and Sister Sledge's "We Are Family" blared over the stadium's sound system. Stargell, Tekulve said, jumped from the bench and called the press box.

"He let them know up there that "We Are Family" was the new team song of the '79 Pirates," Tekulve said. "Lord only knows Willie wasn't the smartest guy in the world, but for some reason he always had incredible timing. They made an announcement that it was the new Pirates clubhouse song, and before you know it, it replaces 'Take Me Out to the Ballgame' for the seventh-inning stretch. Who could have expected the song would give this team an identity, and we'd be forever linked to the song and to Sister Sledge?"

Before his career as a Pirate was terminated with the trade to the Phillies, Tekulve was named an All-Star in 1980, he established the franchise's all-time record for most consecutive appearances as a reliever (722) and for

Kent Tekulve – 119

the most games finished in a single season (67, 1979). His 158 saves in a Pirates uniform was second most to Face's 188, and he paced the National League in appearances four times and games finished on three occasions.

"If I had to say one thing about "Teke," it's that he went out and did his job," said Bill Robinson, an infielder/outfielder with Pittsburgh from 1972-85. "It was never a mystery to him. He went out there and got outs, and he did it in the last inning when, really, he wasn't like most closers in the game."

Face set the major-league record for consecutive wins in relief (18), and consecutive wins (17), in 1959, and on the same day his services were sold to the Tigers on Aug. 31, 1968, Face tied Walter Johnson's big-league mark for most consecutive appearances for the same club with 802. Since his retirement in 1969, Face has witnessed baseball's continued specialization.

"The difference between closers these days varies from one team to the next, but they usually still want the guys who throw real hard and are good for an inning a night," he said. "Tekulve was a big part of it when he pitched. He took more to the mound than I ever did, and his submarine motion was part of it. There's a reason why we haven't seen a lot of guys use that kind of motion, and it's because there aren't too many who have figured out how to make it work for them. Tekulve sure did, though, didn't he?"

Dressed in black and gold from 1975-84, Tekulve averaged 85 appearances and 106 innings. On that last day with the Reds, the sidewinder was 20 appearances shy of Hoyt Wilhelm's appearance record with 1,050 games. He held baseball's all-time mark for consecutive relief appearances – with 1,050 – until southpaw Jesse Orosco surpassed the milestone on June 25, 1999. In 1988 Tekulve, who owned a 94-90 career record with a 2.85 earned-run average in 1,436 1/3 frames, also became the second of three hurlers ever to appear in 1,000 games.

"I knew all the numbers would total something, but it was some time after I retired and got into broadcasting when I realized what the totals represented," Tekulve explained. "I was never the kind of player who paid much attention to the stats until the winter time. That's when I actually had time to digest everything that had happened during the previous season."

The reliever's time away from the ballpark lasted only one season before Tekulve accepted an invitation in 1991 to broadcast five Phillies games once Jim Fergosi moved from the broadcasting booth to manager. The next season, the Philadelphia organization hired the lanky right-hander as a full-time broadcaster.

"I did that, and enjoyed it, for the next seven years until I decided it was time to take some time off again," he said. "You hear it from a lot of older guys – the travel got to me a bit, and I also needed to be home more often for my kids."

However, on April 16, 2001, Tekulve suffered a mild heart attack. "That's why I wasn't able to attend the memorial service for Willie after he died the week before (on April 9, 2001). I was busy recovering from that, but once I got out of the hospital and back on my feet, I decided to get moving again, and that's why I decided to get involved with the Washington (Pennsylvania) WildThings."

The Wild Things, one of 12 teams that compete in the independent Pioneer League, employed Tekulve as their operations director, pitching instructor and first base coach during the 2002-03 seasons. "I enjoyed that opportunity because it let me see the administrative portion of the business of baseball, plus it got me back to coaching kids.

"Those guys were between the ages of 22 and 27, and they were hungry. Because there was no affiliation with

Kent Tekulve – 121

a big-league club, a lot of them saw it as their last chance to get a professional contract – and they played like it. Not only was it fun for me to get back in the game in that capacity, it was also refreshing to see those kinds of attitudes. Those kids reminded me of the way we went after the title in 1979. We all forgot who we were and thought about what we needed to do. That's how you win."

11
Bill Robinson
Pirates Outfielder/Infielder, 1975-82

"One day I would ask him to play the outfield, and he'd say OK. Then the next day, I'd tell him that I needed him to play first base, and he said OK. Bill Robinson didn't care what he needed to do; he just went out on that field and did what he was asked."

— Chuck Tanner

Bill Robinson did not quit winning once he departed Pittsburgh during the 1982 season. The low-toned, seldom-heard McKeesport, Pennsylvania native is, in fact, a four-time world champion. As a player, Robinson played an integral role in the Pirates' title run in 1979, but as a coach he has earned rings as a Metropolitan, a Yankee and as a Fish.

"All year long (in 2003), I asked the same question everyone did - why not the Fish?" said Robinson, the batting coach for the Florida Marlins. "When we broke (spring) camp, I don't think anyone knew what we had, and we did play very well the first couple of months. But then it clicked.

"It wasn't until September when we really started staying close and making a run at it. Our management went out and got Jeff Conine, and Miguel Cabrera and Dontrelle Willis - two kids who were in Double-A that we never really thought of - emerged," he said. "Then all of a sudden, bam."

The Marlins, the National League's wild card qualifier, first defeated the heavily favored San Francisco Giants, and then claimed the NLCS over the cursed Cubs of Chicago. In the World Series, Robinson and the Marlins surprised baseball by ousting the legendary Yankees in six games. "You hear about all the money and the other problems in baseball, but the guys on that team were not in awe of anyone. We laughed, joked and had a lot of fun. It was just a nice period of time."

Robinson was a roving hitting instructor for the Yankees when New York captured the 1999 title with a four-game sweep over Atlanta, and in 1986 he was only a few feet from first baseman Bill Buckner when the Boston infielder committed the most infamous error in World Series history. Robinson was the Mets' first base coach and hitting instructor as New York came all the way back from a three-games-to-two deficit. Buckner's error permitted the Mets to tie the Series at three games apiece before an 8-5 win in Game Seven gave the Mets their second world title since the expansion club entered the majors in 1962.

"That was a great Series, it really was," Robinson recalled. "And really, it's too bad that Bill Buckner is remembered most for making that error when he had a hell of a career (22 seasons, .289 career batting average). What people don't realize is that ball, on the third bounce, had so much topspin on it. It stayed down on him and got under his glove.

"But that happened in the sixth game – we still had to win the seventh for the title. They had as much of a chance to win as we did despite the Buckner error. That was over and forgotten as far as we were concerned, but

Bill Robinson – 125

Boston fans haven't forgotten about it, have they?"

His three titles outside of Pittsburgh have not made Robinson forget his experience with the "FAM-A-LEE" in '79, though, and not just because he was a Pirates player and not a coaching staff member. The outfielder/infielder was not a starter, but he also was not a bench player. "One day I would ask him to play the outfield, and he'd say OK," explained Chuck Tanner, Pittsburgh's manager from 1977-1985. "Then the next day, I'd tell him that I needed him to play first base, and he said OK. Bill Robinson didn't care what he needed to do; he just went out on that field and did what he was asked.

"Winning baseball's world championship is the hardest thing to do in sports for a lot of reasons," Tanner continued. "The number one reason is because you have to have everything come together with all 25 men on your ballclub, and Bill was a big part of that. He was able to play a few different positions without the team losing any offense. You hear about Willie Stargell and the others on the team that got most of the attention, but I don't know if we win the Series that year without Bill Robinson."

The title also proved special to Robinson because of his Western Pennsylvania upbringing. A son of a steelworker, he was raised in Elizabeth, Pennsylvania – just 19 miles from Pittsburgh's downtown district – and graduated from Elizabeth Forward High School in 1961. It was not, however, until Robinson was signed by the Milwaukee Braves and then traded four times before his homecoming took place.

The Braves discovered and signed Robinson in June '61, and the outfielder was then swapped to the Yankees following the 1966 season. The White Sox acquired Robinson in 1970 after his three disappointing years in the Bronx, and Chicago shipped him to Philadelphia in

December 1971. He batted .262 in 306 games and 920 at-bats with 38 home runs and 115 plated runs for Philadelphia, and then was had by Pittsburgh general manager Joe L. Brown in exchange for pitcher Wayne Simpson.

"To come home was great," Robinson said. "The whole 'Local Man Returns Home' storyline meant a lot to me. It got me closer to my father, a man who worked at the Glassport Steel Foundry all his life until it killed him. He worked so hard in those steel mills, and that was the one thing he never wanted me to do …work in a steel mill.

"But for the eight years I played in Pittsburgh, I was my parents' source of entertainment, and that meant so much to me. My father was able to watch me play before he passed away, and that meant so much. A couple of the guys made a 'Mr. Robinson Chair' in our clubhouse and he would come in there after the games. I can remember going in there after the games and looking around for him – he'd be over talking to Willie Stargell or Dave Parker. I'd say, 'Hey, what about me over here?' He'd say, 'Call me tomorrow, I'm busy.' That was my dad's idea of a joke, but those eight years — with him and that team in Pittsburgh — turned my whole life and career around."

Robinson's minor-league career raised expectations, and when the Yankees traded fan-favorite Clete Boyer for the outfielder, Robinson was touted as the second coming of Mickey Mantle. But Robinson batted just .206 in pinstripes, and once the White Sox acquired his services, he was remanded to the minors for the 1970-72 campaigns. "I was also confident in my abilities and knew that some decisions were out of my control," he explained. "I did what I had to do to get back to the big leagues, and eventually I got back.

"By the time I did, I was 29 with the Phillies (in 1972),

and I was playing a couple of different positions and hitting much better than I did with the Yankees. After I came up with Philadelphia, I never went back. When I was traded by the Phillies to Pittsburgh, I was very happy because I didn't have a great taste in my mouth. I had a fantastic year in 1973 – led the team with 25 home runs and really emerged as a big-league star – but in '74, my playing time was cut. I found out years later that it wasn't (manager) Danny Ozark's doing. The front office wanted a young phenom by the name of Mike Anderson to get a lot of the playing time, so I played sparingly."

Pittsburgh skipper Danny Murtaugh immediately defined Robinson's role when the utility player joined the Pirates in 1975, and although he was told he would not be a starter on the NL East-division winning Pittsburgh club, he appreciated the honesty. Robinson would play the outfield when Parker, Richie Zisk or Al Oliver needed rest, and he would pinch hit when necessary.

"Danny Murtaugh looked me in my eye and said, 'You're not going to play much on this club. We're set. We got you as a backup for Richie Zisk. We wanted a veteran player to come in and pick him up and to come off the bench and do some things. We believe you're the man. You make your own breaks from here.'

"Before the end of the year I was playing every day because I deserved to be, and that was fine with me. Earning my way is all I've ever wanted to do and that was possible without the politics. That got my career in Pittsburgh off to a great start."

Robinson's duty expanded each season, but never was he branded with the "regular" label. Instead, he earned the nickname "Super Sub." In 1976, he appeared in 122 games, played five different positions (first base, third base and all three outfield spots), swatted 21 homers, batted .303, was named the team's most valuable player by his teammates and was honored with the Roberto Clemente Award.

But then Murtaugh passed away in December, and with him also died the manager's promise to Robinson. "Danny told me I was going to be the everyday third baseman the next year, but unfortunately Danny died and Chuck came in," he explained. "When Chuck came in he got Phil Garner to play third base. Phil was a second baseman but he played him at third base. So there I was, after having a good year and playing, out in the cold again – just like in Philadelphia. I've always been the kind of person who just wanted the chance. I'll make my own breaks.

"I played to play every day. But from a team aspect, I was more valuable for the team and probably more valuable to myself to be able to play five different positions and do an educated job at all five. That was my job. That's what my manager needed me to do, so that's what I dedicated myself to doing."

Robinson took advantage of the presented opportunities in 1977 as Tanner also used him at the same five spots. He batted .304, led Pittsburgh in home runs (26) and runs batted in (104). The 1978 season saw Robinson's numbers decline (.246, 14 homers, 80 RBI), but his bat returned in full force during the Pirates' magical 1979 campaign.

"Bill was the kind of guy where you go to spring training saying if Bill Robinson has a great year, we're going to have a great year," reliever Kent Tekulve said. "He was just another one of those guys out there doing what he was supposed to do and doing it well."

Robinson played 125 games in the outfield while platooning with left-handed hitter John Milner, and delivered 24 home runs and 75 RBI. In the World Series against Baltimore, the Pirates were behind three games to one following a 9-6 Orioles win in Pittsburgh. But in the next three games, Tanner's team outscored Baltimore, 14-2, to claim the organization's fifth world title.

"I have four World Series rings, but when I look at the Pirates ring I know it's the most memorable for a lot of reasons. I have a lot of memories from that one season. I was on base when Willie Stargell hit the two-run homer (in the sixth inning of Game Seven) and the photo of him and I at home plate ran in every newspaper in the country. I look at that photo sometimes and it gives me the same feeling I felt at the time.

"When Willie hit the home run and came around, I kind of jumped on him, and after the game I remember sitting in my locker crying," Robinson said. " I made a pact with the Good Lord before the Series. I said, 'Lord, if you let us win the World Series, I will not go back to Pittsburgh with the team.' I promised that my wife, Mary Alice, and me would drive home from Baltimore and use that time to praise Him, and to celebrate the victory with Him. And that's exactly what we did because I knew I'd been very fortunate. I wasn't lucky, I was blessed."

Pittsburgh's collection of characters, Robinson said, formed a team that was capable of focusing on the field despite the differences that existed in age, nationality and spiritual beliefs. The glue, he insisted, was the man he personally nicknamed, "Pops."

"Before a game we would argue like brother and sister, but once that umpire said 'play ball,' man, it was like you'd kill for your family. We were brothers. And we had the father figure we needed in the clubhouse, and that's why I started calling him, 'Pops.' If you had a problem, you went to Willie. If you needed advice, you went to him.

"Willie was always there for me and everyone on the team because he wanted one thing – he wanted to win again like he did with Pittsburgh in '71. He tasted it, and once you taste a championship, you want it again and again. That's what he taught me and everyone on that team. I always admired him from afar when I was with the Phillies. When I got there, it was just something I had to say and tell him."

Tekulve explained that although he and Robinson were considered quiet compared to the rest of the roster, it was Robinson's consistency and approach that paid dividends. "He and I were almost similar in one type of way in that we were almost out of place on the '79 team because we were not the loudest or most boisterous.

"We didn't fit into that rowdy disco mold that everybody else fit into. We weren't in that. He was more the quiet professional," the reliever said. "There were a lot of guys around baseball who put up the same numbers as he did, and he was obviously a good hitter. But it was day-in and day-out with Bill, without fail, and he's been teaching the same thing as a coach."

Tanner, Robinson said, handled the group to perfection when arriving in 1977 from Oakland following the wholesale changes orchestrated by A's owner Chuck Finley. The new manager was also a breath of fresh air in his approach to the personalities behind the men in uniform. "The one thing he did – I thought it was just a fantastic move – when he came to the Pirates he did not come and make a bunch of changes. He just kind of blended into us. He set the law, but he blended into the Pirates. There were changes that were made, but he let us go about our business the way we did under Danny and I thought that was as important as anything.

"Chuck also knew all the players personally, and our children and the wives. He'd go up and kiss all the wives, ask you how your family was. I think that's important today. He wasn't that old grumpy kind of guy."

Tanner's optimism was also a key ingredient to the

overall mood of the ballclub, and Robinson appreciated the never-say-die attitude he provoked. "We could be down 15 runs with two outs in the bottom of the ninth, and he would say, 'Come one guys, we got a chance.' He was a never-give-up kind of a guy, and he brought that to the Pirates," he said. "But that didn't mean he was the type of manager who didn't keep guys in line. There was one day when Donny Robinson, who was in his first year, and I had some words. Donny said something smart to me, and Chuck grabbed him by the collar and said, 'You don't ever, ever talk to a veteran like that.' Chuck wasn't all smiles, and the guys who played for him know that."

Robinson appeared in 100 games in 1980 as the Pirates finished third in the East with an 83-79 record. After age and injuries limited his playing time in '81, GM Harding Peterson traded Robinson back to Philadelphia in June 1982. The experience, the utility player explained, is one he wouldn't wish on any player of any era. "When I was informed of the trade, I literally walked down the hallway (of Three Rivers Stadium) to join the Phillies because they were in town when the deal went down.

"It was extremely tough because I had spent eight years with the Pirates and pulling for all of my teammates," Robinson remembered. "Then, one day, I was in the other dugout rooting against them. I didn't like that at all, and I don't recommend that to anyone."

His playing career lasted less than a year following the transaction. In June 1983, the Phillies released Robinson. "I was told that there was a deal pending for me at the time, but I told the Phillies to cancel whatever deal there was, I was done. It was time," he said. "I was 40 years of age, and it was time for me to walk away."

Playing professional baseball was an easy choice for Robinson while growing up near the East Coast's center of industry. His father offered a first-hand account of a steelworker's life, and that understanding assisted him when opting to sign with the Braves instead of agreeing to attend college on scholarship.

"An hour after I received my high school diploma my mom was at home cooking dinner for all these scouts. In those days there wasn't a draft so they were all bidding for me," he explained. "The Braves weren't offering the most money, but they had Hank Aaron who was my idol and I always wanted to play along side of him.

"I think that from age eight I wanted to be a professional baseball player. I never did let anything stand in my way because I never did want to punch a clock. That's why, after I was done playing, I chose to stay in the game coaching and whatever.

Robinson was a hot prospect in the late 1980s and early '90s when interviewing for big-league managerial jobs, and even though he logged time in the minors in Venezuela; Shreveport, Louisiana; Reading, Pennsylvania; and Columbus, Ohio, the call never came. "That's the one thing in baseball that eluded me – I always wanted to manage in the majors. My name did get thrown out there when they were looking for more minorities, but I never got anything, and I'm fine with that.

"How could I possibly complain? My career went full circle, and winning again in 2003 was very special for me and my entire family. After we won the sixth game in New York, my oldest grandson, Brett, was allowed to run around the bases," Robinson continued. "After he was done, he came up to me and said, 'Pap-Pap, this is the happiest day of my life.' I looked down at him and told him that I knew exactly how he felt."

12

Chuck Tanner

Pirates Manager, 1977-85

"Oh, Chuck Tanner smiled a lot. He was smiling unless you gave him reason not to, so you tried not to do that. When you saw Chuck, in the clubhouse or on the field, you wanted to see that smile. If you didn't, you knew something was wrong."

— Bill Robinson

Charles William Tanner played for four big-league clubs, and he managed four teams during the 27 years the New Castle, Pennsylvania native spent in the majors. But since his baseball career veered to scouting in 1989, Tanner has come to be known not for his on-field successes but for his smile.

Those who played for the ever-optimistic Tanner, though, soon discovered how to flip his grin. "Just do something stupid," said catcher Manny Sanguillen, who was first traded to Oakland in exchange for the manager, then played for him from 1978-1980. "He did not smile much when you made a stupid mistake."

Other players from the White Sox, A's, Braves and Pirates also realized how to turn the frown upside down. "The man does like to smile, so that's the easy part," said Kent Tekulve, Tanner's closer every season he skippered the Pirates until the reliever was traded to Philadelphia in April 1985. "All you have to do is what's expected – especially if you made one of Chuck's crazy moves the right move to make.

"Chuck had a side to him that was very much in control. He was the one running the club, and if you had doubts about it you would eventually find out about it every once in a while," Tekulve continued. "You'd get a reminder. Not very often, but you'd get a reminder."

Tanner prided himself on providing equal treatment to each player, no matter his race, nationality, age or level of skill. "I didn't play favorites, and I didn't care if you were a superstar or a bench player. If you were on my roster, I was going to figure how to use you the best so it helped the team the most.

"Hey, I was the boss and I was tough. I smiled – so what?" Tanner said. "There's nothing wrong with taking the best of a situation. I treated people with respect as individuals and encouraged them to excel and I just liked being part of helping my players to achieve their goals. And it was certainly very satisfying when we won – but we won my way. Call me crazy, but I've always been two people – I've always been cocky on the baseball field, but off the field, I'm a different person."

Tanner debuted in 1955 in the Milwaukee Braves outfield after nine minor-league seasons and became the third player ever to hit a homer on the first big-league pitch he saw. His best campaign came two years later when he established career highs in games played (117), batting average (.279), home runs (nine), doubles (19), plated runs (48) and runs scored (47) for the Braves and the Chicago Cubs, but Tanner would never again play with as much frequency.

The Cubs traded him to Boston in March 1959, and the Indians purchased his contract later that season. Two

Chuck Tanner – 137

years and 35 games later, Los Angeles bought his services in 1961 only to use him for 16 plate appearances before his release in '63. "I had one good year, and that was it as a player," Tanner explained. "I could say that I was really never given a chance, but there were a lot of great players in those days.

"I played hard every day and tried to make my breaks. I knew what my calling was, though, and that's why I spared no time once my playing career was over. I wanted to manage, so I went back to the minors to begin my career."

He skippered Quad Cities in 1963-64, and moved to El Paso for the '65-'66 seasons. In 1967, Tanner managed Seattle before returning to El Paso to capture the Pacific Coast League's West Division title. He was hired to manage Hawaii in 1969, and in 1970 the Islanders posted baseball's best winning percentage with a 98-48 mark.

"That's when I finally got my shot at getting back to the majors," Tanner recalled. "That team (in Hawaii) played good baseball so I can't take any of the credit, but if the record is what got me the shot I needed, I don't have any problem with it."

The White Sox hired Tanner to replace Bill Adair near the end of the '70 schedule and he registered a 401-414 mark before leaving for Oakland to manage the A's in 1976. Oakland owner Chuck Finley, however, swapped Tanner for Sanguillen and $100,000 after the ballclub finished second in the American League West with an 87-74 record. The 49-year-old Tanner replaced long-time Pittsburgh manager Danny Murtaugh who passed away in December '76 after battling heart ailments for several years.

The unusual transaction was a welcomed homecoming for Tanner, a three-sport stud and 10-time letter winner in football, basketball and baseball for New Castle's Shenango High School in the early 1940s. "Of all the places I've traveled, I've always loved this part of the country the most because of the people," he said. "The folks in this region are the finest people in the world. They are awesome. They are just so loyal.

"I knew when I came home to manage the Pirates that we would have to win or we would hear about it, and that's because this area has the best sports fans in the world – and they want to win. If we didn't have a competitive team, no one would have cared where I was from, they would have booed and booed a lot."

Instead, Pittsburgh fans cheered, and cheered a lot. Pittsburgh completed the 1977 campaign in the second slot in the NL East, and in '78 Tanner's club waged a heated war for the division title. From Aug. 13 to Sept. 19, the Pirates posted 11-, 10- and seven-game winning streaks and a 29-8 record to diminish the Phillies' lead from eleven-and-a-half games to one. Philadelphia won the title by one-and-a-half games, but it needed 101 victories to do it. "That's where Chuck's optimistic outlook made a difference," explained Nellie Briles, who hurled for the Pirates from 1970-72 before broadcasting Pirates games in 1979-80. "With the way the Phillies were playing that year, every team in the division could have just laid down. But Chuck wouldn't allow that, and that team almost pulled off a miracle."

The experience, Tanner recalled, pulled his club closer together, and following a flurry of moves by general manager Joe O'Toole, the "FAM-A-LEE" was formed. Pitchers Rick Rhoden and Enrique Romo, outfielder Lee Lacy and infielders Tim Foli and Bill Madlock were roster additions before and during the campaign. Suddenly, the "Lumber Company" was transformed into "Lumber and Lightning."

"So many people don't remember that we had to rebuild the team," Tanner said. "We were still the 'Battlin' Buccos' because we never, ever quit, but we also

Chuck Tanner – 139

had the speed we needed, the run producers and some very good pitching.

Pittsburgh began the 1979 season slowly, registering only a 7-11 mark in April. But a 61-30 record in July, August and September allowed the Pirates to outlast the Expos by two games in the NL's East with a 98-64 record. Tanner's lineup featured several league leaders, including leadoff hitter/center fielder Omar Moreno. The speedy Panamanian paced the Senior Circuit with 77 stolen bases, was second with 110 runs scored, and he finished fifth with 196 base hits. Right fielder Dave Parker, the team's only All-Star and Gold Glove winner that season, was the NL leader in extra base hits with 77, second with 327 total bases and third with 45 doubles.

Tekulve led the league with 94 appearances and in games finished with 67, and the submariner's 31 saves were second only to Chicago closer Bruce Sutter's 37.

Pittsburgh scored an average of 4.75 runs per game during the '79 regular season with a .272 team batting average, 148 home runs and 180 stolen bases. The Pirates' pitching staff recorded a 3.41 earned-run average and paced the NL with 52 saves.

"When the season ended and we won a very tough division, we immediately heard how Cincinnati had the better players and that they would go to the World Series," Tanner said. "We didn't necessarily believe anyone who was saying that, but it didn't matter. I knew we had the better team, and after we swept the Reds in the playoffs, we thought the same thing going against Baltimore.

"Everyone was saying that Earl Weaver was a great manager, and that he's this and he's that. Hey, Earl was a great manager, but there are a lot of things that go into winning. I was once at a banquet with Earl Weaver, and I told him how he had the best players, but I had the best team. And I also told him that the Pittsburgh Pirates had one other advantage over his team – the Pirates had the better manager. Weaver began screaming about how he was going to kill me."

The Orioles won 102 games in 1979, capturing the AL East Division by eight games over Milwaukee while boasting a lineup featuring catcher Rick Dempsey, first baseman Eddie Murray and outfielder Ken Singleton. Weaver's rotation possessed 23-game winner Mike Flanagan, Dennis Martinez, Mike McGregor and Jim Palmer.

Flanagan notched a complete-game victory in Game One of the Series, and McGregor followed suit in the third contest with a nine-inning, nine-hitter. After Orioles reliever Tim Stoddard registered the win in Game Four, Baltimore owned a 3-1 advantage.

"They had all those guys, and we didn't even have a 15-game winner in our rotation or a 100-RBI guy in our lineup," Tanner said. "How did we ever survive, let alone win the whole damn thing? It had never been done before and it hasn't been done since, but I'll tell you how we did it – we were that 'FAM-A-LEE' that everyone was talking about."

Tanner's club outscored the Orioles, 15-2, in the final three games, Tekulve nailed down a pair of saves, and first baseman Willie Stargell delivered the key blow in the sixth inning of Game Seven. "McGregor was pitching a great game and he was ahead, 1-0. But then my man, Willie Stargell, went up there and came up big the way superstars do," Tanner explained. "Willie hit the home run we needed to put us over the top, and we all came together to get the job done because we had to. No one in our dugout wanted to lose that Series.

"We had that song ("We Are Family") by Sister Sledge and everyone sang along during the seventh-inning stretch all season long. That was great, but the best thing was that it was true. We were a family and that's why we won."

Stargell, of course, was the club's catalyst. He was 39 years old, a 17-season veteran, a seven-time All-Star, and known as "Pops" in 1979. By the end of the schedule, Stargell had cranked 32 regular-season homers to earn a tie with the Cardinals' Keith Hernandez as the NL's MVP, was named the most valuable player in the Pirates' sweep of the Reds in the NLCS, and was the World Series MVP, as well. "By that time, Willie was getting older so I didn't play him all of the time. I rested him when I could so I could keep him fresh because I knew we would need him in the end," Tanner explained.

"But he had the makeup of a ballplayer, and the other guys listened to every word he said – and that was a great thing for our team because Willie was baseball. He listened and learned his entire career from some of the greatest players in the world. He took it all in, and then he shared it. That's what made Willie so special. That's what made him a crown jewel."

Tanner's smile was never so broad as after Pittsburgh returned from Baltimore toting the game's World Series trophy, the franchise's fifth world title and third since 1960. "Chuck seldom takes any credit," Tekulve said. "But he was a very good manager – for that club in particular. It was a veteran club, and what Chuck did best was he put us together, watched over us and just let us play. It was never important for him if anybody knew who was pushing the right buttons or did the right thing. One of his favorite lines was, 'You guys take credit for all the wins, I'll take the blame for the losses.'

"We showed up every day and did our jobs. We owed it to our teammates, played the game as it was supposed to be played and then we went home and got ready for the next one. That was Chuck's program and it worked."

"Oh, Chuck Tanner smiled a lot," outfielder/infielder Bill Robinson recalled. "He was smiling unless you gave him reason not to so you tried not to do that. When you saw Chuck, in the clubhouse or on the field, you wanted to see that smile. If you didn't, you knew something was wrong.

"Chuck let us be us as long as the job was getting done on the field, and that's what good managers do, in my opinion," Robinson continued. "He deserves a lot of credit, even if he doesn't want it."

Pirates fans continued cheering in the years that followed the world title, as Pittsburgh remained competitive despite the retirement of Stargell and the departures of several key players. The Pirates fell to third in 1980, struggled during baseball's split season in 1981 with an overall 46-56 record, and was fourth in '82 with an 84-78 mark. In 1983, Philadelphia – powered by Mike Schmidt, Pete Rose and Joe Morgan and the pitching of southpaw Steve Carlton – won the East by six games over Pittsburgh.

The cheers then turned to jeers. Although Stargell joined Tanner's coaching staff, Parker left the Pirates via free agency, and suddenly the team's daily lineup featured a youth movement of no-names and a core of veteran has-beens. "Money got tight – real tight," Tanner explained. "No one can blame Dave Parker for leaving Pittsburgh because he had the chance to earn more in another city. Every other player around baseball was doing the same thing.

"The Pirates paid him a million dollars, and good for him, but he kept getting better and could get more. I told him I understood it, but the fans didn't like it. Every time he came back to Pittsburgh I'd give him a hug, but the fans really let him have it. That made me mad every time."

In 1985, the jeers turned to sneers. Not only were the Pirates the last-place finishers in the NL East, but drug abuse in the big leagues arrived to the forefront in Pittsburgh. Curtis Strong, a caterer for the Phillies, was charged with 16 counts of distributing cocaine, and several major-league players were counted among his customers. What made matters worse was that the proceedings were in federal court in Pittsburgh.

Parker, John Milner and Dale Berra joined Tanner in testifying during the 12-day trial, and several former Pirates were implicated – including Madlock, Stargell, Lacy and relief pitcher Rod Scurry. Parker, Berra and Milner all admitted to purchasing cocaine from Strong between 1980-83, and Strong was found guilty of 11 counts of cocaine distribution on Sept. 20, 1985. He was sentenced to 165 years in prison and fined $275,000 by Federal Judge Gustave Diamond.

"The problem wasn't just in Pittsburgh, but that's where they had the trials so we suffered the most," Tanner remembered. "Because the trials were in Pittsburgh, everyone believed that it was the Pittsburgh Pirates who were the problem and that just wasn't true. There were a lot of teams out there who were experiencing the same problems we were, but that year was the toughest year I ever had managing because drugs ran rampant.

"I know now that I had a few guys using drugs, and that I didn't know it. Hell, I didn't know the difference between cocaine and salt. But Chuck Noll over at the Steelers had it, and he didn't know it. It was in the NBA, the NFL, Major League Baseball and the entire country. It was the era when, if somebody wanted drugs someone else could easily go get it for them."

The 1985 season, in fact, was Tanner's last as manager in Pittsburgh even though he had two more years on his contract and an above-.500 record. The Galbreath family, the ballclub's owners since August 1946, decided to sell the franchise to the Pittsburgh Associates, a civic grouping of Pittsburgh business executives, and one of the first decisions made was to replace Tanner with Jim

Leyland. "The Galbreaths were great owners. I think they got out because they were tired of losing money," Tanner said. "I had a meeting with the Galbreaths and the new owners at Three Rivers Stadium, and they told me they were going to make a change. I told them it was their loss."

Tanner, who, in nine years, collected a 711-685 record as the Bucs' skipper, was soon hired by Atlanta to replace Bobby Wine and guide a ballclub that finished the '85 campaign at 66-96. With Stargell at his side, the Braves failed to improve, however, posting 72-89 and 69-92 records in 1986-87, respectively. When Atlanta sat at 12-27 in 1988, Tanner was fired.

"I've heard that people believe I had a tough break in Atlanta," Tanner explained. "But I don't think so. Ted Turner was great to me because he knew how much work that organization needed when he hired me. And we got a lot of that done before that change was made. Hey, I've been lucky to have had great owners everywhere I managed. Chuck Finley was good to me in Oakland. John Allen, who owned the White Sox, was good to me. And the Galbreaths were the best owners in the world. They were all loyal to me, and that's why I gave that back to them.

"I had chances to manage other clubs, but I didn't change jobs because I was faithful to the people who gave me my jobs in the majors. I told every one of them that there are only two things I know about baseball – loyalty and ability. The Red Sox called when I was managing the White Sox, and I turned them down. The Yankees called when I was managing the Pirates, and I turned them down even though the Galbreaths told me to go and double my money," he continued. "Immediately, I told the Galbreaths that as long as they were the owners and they wanted me to be the manager, I'd be there."

Tanner, who fathered Pirates bullpen coach, Bruce Tanner, managed 2,738 major-league games and completed his 19-year career with a 1,352-1,381 record and .495 winning percentage. His ballclubs finished sixth on five occasions, in fifth and fourth place three times each, third twice and in second place after five regular seasons.

One season, though, Tanner lived his dream. "I've never worried about the records, I've worried about getting the best out of my players. I pushed them to make the progress I thought they could make ... and it didn't matter if we were in the running for the division."

"What makes good managers are good players," Tanner insisted. "Look at (Yankees manager) Joe Torre. You have to have good players, and you have to cope with them, manage them and communicate with them."

Tanner, a prostate cancer survivor and victor over heart by-pass surgery in 1992, turned to scouting immediately after his Atlanta tenure ended, and he served as a special assistant to the GM for the Brewers for 11 years before Cleveland hired him in 2003. He has, during a career spanning nearly seven decades, offered his opinions, philosophies and strategic thought process. Along the way, however, Tanner succeeded in becoming Pittsburgh's version of the Yankees' great-for-a-quote Yogi Berra:

> "It's hard to win the pennant, but it's harder to lose one."

> "The greatest feeling in the world is to win a major-league game. The second greatest feeling is to lose a major-league game."

> "There are three secrets to managing. The first secret is to have patience. The second is to be patient. And the third most important secret is patience."

"You can have money piled to the ceiling, but the size of your funeral is still going to depend on the weather."

Tanner explained, with shrugged shoulders and a broad smile, "What makes sense makes sense. I've always said what has been on my mind and had fun with it. But when it's come to baseball, I'm sure everything I said was on the mark.

"It's always been baseball ever since I can remember, and I know I can never repay the game for what's it's given me. It's all I ever wanted. I never wanted anything else. I didn't want a boat. I didn't want an airplane. I didn't want to go on vacation. I didn't want anything besides baseball, and that's why my wife (Barbara) has said that I've never taken her on vacation. I've told her that I took her to Hawaii, and she said, 'Yeah, for the baseball meetings.' I told her that I took her to Dallas, and she said, 'Yeah, for the baseball meetings.' Trust me, she knows how I am, and she knows baseball is the greatest game in the world."

Tanner did not receive many Hall of Fame candidacy mentions during the 15 seasons that followed his managing career, and his record, he said, does not demand consideration. "My goal is to be in the Pirates' Hall of Fame," Tanner admitted. "That would be better than anything. It would be better than being in the Hall of Fame in Cooperstown. That would make my whole career perfect."

13

Barry Bonds
Pirates Outfielder, 1986-1992

"Barry did treat a lot of people like crap, but he was young and immature. At the same time, Barry did a lot of things that weren't written about in print. He did a lot of charity work that was never heard about, and I think it's a little unfair. He's a different man now than he was then. I think we all are because we all grow up and we all put childish things away, and he had to put his childish things away."

— Lloyd McClendon

Two California men battled for more than a year for legal possession of a baseball, one smudged by the pre-game "rubbing down" ritual and dented by the bat of Barry Bonds on the final day of the slugger's record-setting 2001 season. The ball was the seventy-third sent over big league walls that year by Bonds, a count that eclipsed Mark McGwire's total of 70 home runs the former Cardinal swatted in 1998.

McGwire's ball fetched a multi-million-dollar bid by comic book creator Todd McFarlane. Restaurateur Alex

Popov and software engineer Patrick Hayashi envisioned a similar payday when they advanced the dispute over the Bonds ball to a three-week court battle. Popov claimed he caught the homer but was mugged and bitten by Hayashi in the melee that ensued after Bonds sent Dennis Springer's pitch over Pac Bell Park's right field wall. The trial ended with Judge Patrick McCarthy ordering the same resolution Bonds suggested moments after banging the baseball into the game's record books – sell and split. McFarlane bought the ball, ironically, for only $450,000.

But why the fight in the first place? Is not Bonds, despite a long list of splendid accomplishments compiled over the first 18 years of his major-league career, the same man who had been accused of being selfish, arrogant, a clubhouse cancer and America's least loved superstar even after establishing a new standard for power hitting? In Pittsburgh, Bonds was jeered consistently – during each plate appearance and on every occasion he touched the ball on defense – after signing with San Francisco in 1993, and his former manager was angered each time.

"When the Pittsburgh fans have booed Barry, it's hurt me because this guy did nothing but bust his tail for the Pittsburgh Pirates," said Jim Leyland, the Pirates' skipper from 1986-1996. "He played great for the Pittsburgh Pirates. He was a young player who maybe wasn't ready to handle all the pressures that went along with that time, but that's a thing of maturity.

"But what bothers me is that I will guarantee you a lot of the people who are booing would ask Barry to sign an autograph if they had a chance. If Barry had an autograph session at PNC Park, the same people who boo him would get in line. I believe that with all my heart, and it upsets me. He was a tremendous player for the Pirates, and that's what the fans should remember."

Barry Bonds – 151

Bonds first appeared in a Pittsburgh uniform in 1986 as a center fielder after just one minor-league season. By 1992, the left fielder's rookie salary of $60,000 swelled to $4.8 million after he captured the 1990 NL MVP Award, two Gold Gloves and an All-Star appearance. But Bonds' aloof approach with the city's press corps, and his claims that he often was misquoted by media members, strained his relationship with the public and may have cost the outfielder commendations beyond what he was awarded.

"Barry liked to use the 'misquoted' excuse as a cover whenever he said something outspoken but later realized it was too late to take it back," explained John Perrotto, a Pirates beat writer for the *Beaver County Times*. "Barry could go to both extremes with the media. He could be extremely moody and unapproachable. Other times, he could be very charming and glib.

"He sometimes eased the standoffish part. He would never talk about his father and would get angry if you ever brought Bobby's name up. In fact, a lot of times, reporters would get mixed up and call Barry, 'Bobby,' and that would set him off," Perrotto continued. "But in August of 1990, Barry was on the verge of (collecting 30 home runs and 30 stolen bases) for the first time and his father had gone 30-30 five times in his career. When I broached the subject, Barry was great and we talked for 45 minutes about his relationship with his father and how he resented him as a kid because he was always away from home. He was outstanding."

Bonds' strained relationship with teammates, namely former San Francisco second baseman Jeff Kent, received much coverage, as well, but not all those who shared locker rooms with the California native experienced what has been described as a self-centered nature. "Barry and I bonded very quickly when I came to the Pirates," said Lloyd McClendon, a utility player for

Pittsburgh from 1990-1994. "I had heard this and that about him, but I've always been a person who gives someone a chance. When I met him, I looked him in the eyes and told him to go screw himself. That made him the happiest person in the world, and we've been friends since.

"Yeah, Barry did treat a lot of people like crap, but he was young and immature. At the same time, Barry did a lot of things that weren't written about in print. He did a lot of charity work that was never heard about, and I think it's a little unfair," McClendon continued. "He's a different man now than he was then. I think we all are because we all grow up and we all put childish things away, and he had to put his childish things away."

The outfielder batted .301 with 33 homers, 114 plated runs and a career-high 52 stolen bases in 1990 and was named the National League's most valuable player after leading Pittsburgh into the postseason for the first time since the Pirates won the world title in 1979. In 1991, he hit at a .292 pace with 25 home runs, 116 RBI and 43 base swipes, but he was the runner-up to St. Louis' Terry Pendleton in the balloting. "Sure, Barry made some mistakes with the media, but we all do," said Leyland, who skippered the Marlins to the World Series title in 1997 and also managed in Colorado. "I really believed not winning the MVP that year hurt Barry Bonds. I think the media that year was upset with him, and I really believe that's why he didn't get it because he really should have won it.

"Barry believed that as long as he got his job done on the field, there was no need for him to sit and explain it to anybody. And I know Barry was brash at times, but so were a lot of guys in the clubhouse. I know I was," Leyland continued. "But we weren't worth the ink because we weren't superstars. When you're dealing with superstars, it's totally different than dealing with somebody else. Is that fair? No. Is that the way it is? Unfortunately, yes."

An incident that occurred during the ballclub's spring session in March 1991 tarnished Bonds' reputation with Pittsburgh fans. Although the coaches involved later explained the confrontation between the left fielder and Leyland as a misunderstanding, national media was on the grounds of Pirate City with cameras rolling and microphones recording.

"I've been kissing your butt for three years," Leyland said in the direction of Bonds. "If guys don't want to be here, aren't happy with the money they're making, don't take I out on everybody else."

"What the fans saw and heard on TV was misleading," said Bill Virdon, a former player, manager and bench coach for Pittsburgh who was a coach on Leyland's staff at the time. "Barry had a guest on the field with him that day, and he was told that the guest was not welcome by someone in the organization. That happened just as we were getting the outfielders together for drills.

"When Barry joined the group he was still hollering about his friend being asked to leave the field. I said, 'Barry, knock it off.' He said, 'Nobody's going to tell me what to do,' but he was talking about that guy getting run off the field, not about what I said to him. That's how that started. So I jumped on him and Leyland saw this happening because he just so happened to be walking by. He walked in and took over and I got out of it. But after that, I never had any problems with him or with instructing him. We've always gotten along very well."

Although Leyland refused to regret his reaction and words to Bonds, the media play the incident received was unjust in his opinion. "At the time, there were some other things in my head – Barry was never good about cooperating for team photos and things like that for whatever reason," Leyland said. "So when I heard what

Barry said to Bill, I thought he was being disrespectful and I lost it. The only difference between that situation with Barry and others I had with other players was that it was on camera.

"The only regret I have about that incident is that I think people kind of accepted me more because they felt like I got on some star player. That certainly wasn't the point. I have five brothers and sisters. We had fights at home and we argued. I just hope the baseball fans in Pittsburgh accepted me for being a good manager, not the fact that I had a disagreement with Barry Bonds. People still come up to me and say, 'Boy, I loved it that one time you told off Barry Bonds.' I don't want to be known for that incident because that's not important to me."

Bonds is often criticized for his postseason performances while with the Pirates. In Pittsburgh's National League Championship Series loss to the Reds in 1990, he supplied a .167 batting average with just three hits in 18 trips to the plate. During the '91 and '92 NLCS defeats to Atlanta, Bonds compiled a .200 batting average with one home run and two RBI in 50 at-bats. What made matters worse were two facts – the outfielder was destined to depart Pittsburgh following the 1992 campaign because the franchise was financially incapable of retaining the future free agent; and because his last act in a black and gold uniform was his failure to throw out the slow-footed Sid Bream as the Braves completed their last-inning, Game Seven comeback to advance to the World Series.

Bonds and the Pirates were down three-games-to-one in the series, but after victories in games five and six, masterful right-hander Doug Drabek took a 2-0 lead into the final game's last inning. After the Braves slimmed Pittsburgh's lead to 2-1, Leyland removed Drabek in favor of right-handed reliever Stan Belinda. With two outs, Francisco Cabrera – who spent most of the season in the minors – punched a pinch-single to Bonds in left that scored David Justice with the tying run. Bream then slid safely past diving catcher Mike LaValliere. It was the first time in postseason history that a team went from losing to winning on the final pitch of the decisive game.

"To this day, I still think Barry's throw was a hell of a throw," McClendon recalled. "Barry really went off line and made a pretty damn good play at the plate. It just wasn't for us that day, and it was for the Braves. It was the toughest defeat on my life, but in no way do I blame Barry for that loss, or any of them."

"I know Barry has been criticized for not throwing Bream out," Leyland said. "But he made a great throw. He had to go to his left a little bit to get the ball and he's a left-handed thrower. That's a tough play – not many outfielders could have made it as close as it was.

"There are a lot of fans who have singled him out for not hitting in the playoffs, but there were a lot of guys who didn't hit in the playoffs. That wasn't fair. Jay Bell batted .172. Jeff King hit .241. What about those guys?" he said. "After that loss (in Game Seven), I was in a fog. I felt terrible. I just kind of mumbled a few things to the team in general because I knew how bad they felt and I knew how bad of a heartbreaker it was."

Outfielder Bobby Bonilla, a member of the 1990 and '91 division-winning clubs, left for the New York Mets before the 1992 season to become baseball's highest-paid player, and 20-game winner John Smiley was traded to Minnesota. Yet, the Pirates registered a 96-66 record to claim the East crown and return to the playoffs. The loss to the Braves, however, represented the team's final opportunity before Bonds and Drabek would follow Bonilla out of town to teams with deeper pockets.

"After the loss to the Braves in the NLCS," Perrotto said, "I've never ever – even on the high school level –

seen more players crying after an athletic event than on that night. They closed the clubhouse for about a half-hour and guys were still devastated once they let us in. Barry talked after every game in the postseasons back then, but even he was speechless and didn't want to talk.

"Before the media was allowed into the clubhouse after the loss, Leyland came out and said, 'Fellas, I know you're all on deadline, but can you please give us a few more minutes.' He was as white as a ghost. He looked like he had just watched someone die – and in a way, he did. He watched three years of frustration come to a head and knew it was the end of an era because they weren't going to win anytime soon. In '90, it was a case of just being happy to be there. In '91, it was frustration because they knew they had the best team in the majors. In '92, it was beyond frustration because they knew that was their last chance."

The organization's refusal to compete for Bonilla's services following the '91 campaign, and the trade of Smiley to the Twins, sent a message to Leyland and those who remained on his roster. "I think everyone knew it was 'now or never,'" the manager recalled. "We knew it was going to be a start-over process, basically. It was sad because I remember when I came to the Pirates I said, 'We don't want to just have a good year at some point. We want to be known as an organization to have a competitive product, year-in and year-out.' And with that group of guys we would have had that. We would have been good for a long time. That team had a chance to be good for a long time, but you could see the hand-writing on the wall.

"It was hard to say goodbye to those players because they were my babies, so to speak. I loved those guys, and they knew it. We went through a lot together and had a lot of great moments, but we'll never really get recognized and rightfully so. In '91 we had the best team in baseball and we didn't win it. In '92 everyone was disappointed. We came back and we could have gotten there in '92, but we just didn't quite get over the hump. But that doesn't take away from the way you felt about your players or the way they felt about each other. It was great group. We grew up together and turned the franchise around. All those guys – Mike LaValliere, Andy Van Slyke, Barry Bonds, Bobby Bonilla, Doug Drabek, John Smiley, Jeff King and Don Slaught – had to leave and each time it was tough."

Bonds, however, did not wish to leave Pittsburgh. The Gold Glover, in fact, was receptive to negotiating a multi-year agreement that would have proven to be the game's richest contract but one worth less than what was offered elsewhere. "He said five years for $25 million," explained Perrotto. "I ran that by (general manager) Ted Simmons and he was agreeable. However, the deal died when it went a step further, and Barry wound up getting six years and $43.5 million from San Francisco that winter. If the Pirates would have listened to him, though, they would have had Barry at a heck of a bargain."

"I really don't think Barry wanted to leave Pittsburgh," Leyland said. "I believe that with all my heart. But I think when we really didn't make that effort to get him signed, he kind of felt like he had to defend himself for leaving, and I don't think he did. I think that's also where he got himself in a little bit of trouble with the Pittsburgh people. He felt like he had to find something wrong with Pittsburgh to justify leaving. In reality, the fact was $18 million is a pretty good reason to leave and the man should have had, and did have, the right to exercise his chance to get a pot of gold at the end of the rainbow."

And yet, Leyland and Perrotto continued to hear the boos in 2003 when San Francisco visited PNC Park more than a decade after Bonds went West. "He doesn't deserve it for a lot of reasons, but mostly because he didn't want to leave the Pirates," Perrotto insisted. "Keep in mind, Barry bought a house in Moon Township and lived here in the offseason.

"I think the media has learned to respect him as he's gotten older. I think the fans have, too, except in Pittsburgh. People here just can't let it go, even though it's been more than 11 years since he left. Why? Honestly, the fans can only answer that question. I don't understand it. I don't have an answer."

Bonds became a Giant, literally and figuratively, when his career continued beyond his years with the Pirates, and not only did he set the new record for home runs in a single season but he also became the first and only member of baseball's 400-400 and 500-500 clubs for homers and stolen bases. San Francisco has suffered just two losing seasons with Bonds on the roster, and the Giants have won four divisions and appeared in the 2002 World Series against Anaheim.

By 2003, Bonds had been named to his twelfth All-Star squad, had won his unprecedented sixth MVP Award to become only the fourth player in baseball, football, basketball and hockey to win the honor six times, possessed eight Gold Gloves and had earned more than $110 million playing the game.

By joining San Francisco's ballclub, Bonds also became a member of the team for which his father, Bobby, and his godfather, Willie Mays, excelled. Barry's pedigree was heralded when Pittsburgh selected the sleek outfielder in the first round of the 1985 amateur draft out of Arizona State University, and he soon placed on display why he rightfully deserved to be the sixth player selected overall.

"You could see his talent right away," Virdon recalled. "He swung the bat very well, and he had speed. The one thing that he lacked was a strong arm, but he worked hard so that didn't hurt him. He learned to charge the ball and get as close to the infield as he possibly could. Nobody ran on him because he did what he had to do to keep them from running on him. You have to take your hat off to him for that."

Although Leyland admitted he never imagined Bonds would connect for 73 home runs in a season, he was aware of the player's potential. "I saw a very talented young player who was a little bit beyond his age really in baseball ability," Leyland said. "He had tremendous instincts. He played older than he was. You saw the definite makings of a superstar, and you knew it was just going to be a matter of time. It was interesting breaking him in and being his friend and watching what he went through, but you knew at some point you were going to have a superstar on your hands. There was no question in that."

"He had it all," Perrotto said. "And you knew it right away. I covered his first game, and I will always remember how dejected Barry was about going 0-for-5 in his debut. It was like the end of the world. He just sat there with his head down at his locker like he was the world's biggest failure."

With the Pirates, Bonds averaged 25 home runs each year for seven seasons, but with San Francisco he increased that statistic to 44 per. He remained healthy for the most part, playing in at least 130 games except in 1994 and 1999 due to injuries. And despite living through the death of his father in August 2003, Bonds collected his eleventh Silver Slugger Award as the NL's top power hitter with a .341 batting average and 45 homers. He has said, however, that his goals on the diamond were not determined by what past players had accomplished, but instead on the feats of his father and Mays.

Bobby Bonds and Mays employed the rare mixture of

power and speed during their respective performances. Mays, a 12-time Gold Glover and two-time MVP, was inducted into baseball's Cooperstown shrine in 1979 after appearing in 20 All-Star games in 22 years. Bobby Bonds was an All-Star on three occasions, won three Gold Gloves and played in seven cities for eight teams in 14 years and was a 300-300 club member upon his passing.

"Obviously his dad was a great player," Leyland said. "There's no question about that. Plus, I think being around it so much like Barry was at a young age was very beneficial. He saw at a very young age what it takes, and I'm sure his dad talked to him about all that stuff.

"Barry certainly had an advantage, but he was his own man. He made himself a great player. Having his father's and Willie's careers to measure himself against was both good and bad – most of all, I think those men motivated him to play above and beyond what they did on the field. I think it motivated him to meet those goals he had and to be that kind of player, and certainly he's done that."

"I think Barry has played haunted more by his godfather's career than by his dad's," McClendon said. "I don't think he and his dad had the best relationship. I think Barry grew to love his father more as he matured because he came to realize how precious time really is."

Bonds announced his intentions to become "nicer" to media and fans alike prior to the start of the 2000 season, a decision, McClendon believed, was made because the superstar finally started feeling the effects of the battle all athletes wage against age. "One day Barry probably woke up and felt something he'd never felt before," McClendon explained. "He has always kept himself in tremendous shape, but when you get older in this game you start feeling new pains every once in a while. They remind you that the end is inevitable.

"Barry is likely worried about his legacy and how people will look at him when his career is over. That's only normal," McClendon added. "I know he's said that he only wanted the respect of his father and godfather, but I believe he wants to be respected – maybe not as a nice person, but as one of the greatest baseball players to ever play the game. If you look at just his numbers, and not at what he's said during his career, you can't help but respect him. He's a first-ballot Hall of Famer, for sure."

"Barry deserves the respect," Virdon said. "The question is, will he ever let people respect him? Personalities are different. I just don't think he likes all that attention. I think it bothers him and he'd rather not have it, and his reactions are not treated favorably because of who he is."

Bonds raced to the forefront as the player with the best chance to surpass Babe Ruth's homer total of 714 and Hank Aaron's all-time record of 755 with his 73-home run season in 2001. He ended the 2003 campaign just two short of his godfather's total of 660 with three years remaining on the five-season, $90 million contract he signed before the '02 schedule. Leyland said, though, that even if Bonds retired without reaching Mays, his former left fielder should be considered the greatest player ever.

"I think if people don't let their feelings get in the way and look realistically at it, that a lot of people would have to say he may be the greatest of all time. I think he is, and that's my opinion. I'm sure there are a lot of people who say Willie Mays was better," he said. "I really believe a lot of the fans, even the ones who boo him at PNC Park, are truly wondering if they are watching the best ever to play the game.

"I know he's the greatest player I ever managed, and

honestly, it's an honor to say that I managed him. That's why I think there's a great chance that he may go down as the greatest player of all time."

14

Lloyd McClendon

Pirates Utility Player, 1990-94
Manager, 2001 - Present

"Lloyd respects the game of baseball with all his heart and soul, and that's something you don't find in players or managers these days. It's too much about the money – but not with Lloyd. To him, it's about winning, and if he's not winning, then it's about finding ways to win."

— Bill Virdon

Lloyd McClendon was a big-kid bully at the 1971 Little League World Series in Williamsport, Pennsylvania. He was much larger than most 12-year-olds, already standing at 5-foot-10-inches tall and tipping the scales at 170 pounds, and was the Anderson All-Star's best pitcher when not catching.

McClendon was also the superior hitter on the club that hailed from his hometown of Gary, Indiana, a city founded and sustained by the United States Steel Corporation. Gary thrived before McClendon's 1959 birth, but it was a depressed community while his parents raised him and his 12 brothers and sisters. "We all grew up in the inner city of Gary, so I think it's easy to say that I wasn't raised with a silver spoon in my mouth," McClendon said. "We had a lot of hard times, but if I can only be half the man my dad was I'd consider myself successful in life. He raised 13 kids, and somehow none of us went without anything we needed in life. Certainly, we didn't get everything we wanted, but we had clothes on our backs and there was food on the table.

"My father made parts for buses on an assembly line, and my mother was home with the kids, so they both worked harder than hell. And I know I didn't get a lot of new things – I had hand-me-downs all the way – but I don't remember caring much at the time. Today, I'm proud as hell of my parents, and I hope they're able to look down on me now and think they raised a decent man."

The Anderson All-Stars were the pride of Gary when the team advanced all the way to Little League Baseball's final tournament in central Pennsylvania. The club represented the city's best 11- and 12-year-olds selected from eight teams. "And we were all the right ages, and we were all from the city of Gary," McClendon explained. "We played pretty well with a little talent and a lot of luck, but I don't think anyone realized what we had accomplished until a couple of years later, to be honest. We were kids and we were playing baseball – that's all we knew.

"And I wasn't the biggest kid on team," he said with a grin. "I think I was only the fourth biggest kid on the team, and there were teams with kids even bigger than us. It's the age – some kids are bigger than others. You didn't hear me making any apologies because I had my share of home runs that year. Over the course of the season I probably hit 20 home runs."

Lloyd McClendon – 167

Once arriving to Williamsport, McClendon supplied a power performance still not forgotten several decades later. He was Anderson's clean-up hitter, and following the team's first two victories, McClendon owned four home runs in four at-bats, and he was also intentionally walked on a few occasions.

"I wasn't too happy that I was getting walked because I wanted to hit the damn ball. I was a young kid who really didn't understand what the heck was going on. You're 12 years old and you want to play the game," he said. "Our first two wins got us into the championship against the team from Taiwan, and during the press conference before the game their coach was asked if they were going to walk me and he said no."

McClendon, however, swatted his fifth homer in his first at-bat of the game. "And they walked me every time after that. Before the game, the coach said they would rather lose and go home proud than resort to walking me, but that didn't happen."

The Andersons did not win the title that August, losing to the Taiwan club, 12-3, with McClendon's single swing supplying the only three runs scored against the Far East representative in the tournament. McClendon also pitched the championship game, whiffing 15 batters in eight innings. "Yeah, well, their pitcher struck out 22 of our guys and all we saw was them celebrating on the field after the game," he said. "But it was still a dream to live and it's one of those memories I'll probably never forget.

"Someone tagged me with the 'Legendary Lloyd' nickname some time after that World Series and it has stuck every since," he continued. "But you know how a story changes over the years – it wasn't five home runs on five pitches, it was five homers in five at-bats and 10 intentional walks."

But while the legend grew large, the player did not.

McClendon's height increased by only an inch while playing football, basketball and baseball for Roosevelt High School. Although he was the first scholastic performer ever to make the school's varsity baseball team as a freshman and was named All-Indiana following an impressive senior season, "Legendary Lloyd" attracted few scouts and was not drafted. McClendon opted to attend Valparaiso University. "In high school, nobody called me 'Legendary Lloyd.' Some teams looked at me, but I did not get drafted until college. I weighed 195 pounds and was 5-foot-11. I had decent speed, hit to all fields and had an average arm. I was an average defensive catcher but an offensive player.

"It was always my aspiration to play pro ball. I always remember my educators asking me what I wanted to do when I grew up. Some people would say they wanted to be an accountant or a scientist. My answer was the same every time – I wanted to be a baseball player."

After batting .330 with 18 home runs and 73 runs batted in over three seasons at Valparaiso, McClendon was selected by the Mets in the eighth round of the 1980 amateur draft and immediately assigned to their rookie club in Kingsport, Tennessee. During his seven seasons in the minors, McClendon experienced the business of baseball, injuries and racism.

"It was tough to be a black man in the minor leagues at the time, especially when you were assigned to the small, southern towns," he recalled. "I remember Darryl (Strawberry) and I getting chased one time because of the color of our skin. And there was a time when I couldn't find an apartment to rent. I went to this one place and was told that it had been rented. The next day, one of my teammates said they rented the apartment after I had visited the place. He was white.

"My dad taught me that if you want to change things, complaining is not the way to go about it. He told me to get on the inside, so I called the Mets and told them about the trouble. To the Mets' credit, the situation improved."

New York traded McClendon to the Reds following the 1982 season in a transaction that returned Hall of Famer Tom Seaver to the Mets. McClendon climbed Cincinnati's farm system ladder until finally making his big-league debut in 1987 at the age of 28. "When I got traded to the Cincinnati Reds, I thought my career was really going to take off, but then I broke my wrist and I had some shoulder problems. When I finally got healthy, I started putting up some good numbers and got the call I'd been waiting for all my life."

McClendon played first and third base, in the outfield and behind the plate for Cincinnati manager Pete Rose in 1987-88, but his offensive production (.208, .219) banished him to utility status. With his career on the line, McClendon welcomed the move made by Reds' general manager Murray Cook when Cincinnati traded the Indiana native to the Chicago Cubs.

"We had a very talented club in Cincinnati so I talked with Pete (Rose) at the end of the season," McClendon explained. "I said, 'Could you do me a favor? I love playing here, but I'd like you to trade me because I want to play.' He took care of me."

The Cubs' roster possessed catchers Damon Berryhill and Joe Girardi, so McClendon was asked by manager Don Zimmer to play more outfield than anything once he was recalled from Triple-A Iowa in June. He also spelled Mark Grace at first base and Vance Law at third while batting .286 in 92 games with career highs in home runs (12) and RBI (40). The Cubs, meanwhile, claimed the National League East Division by six games over the Mets with a 93-69 mark.

"I still wasn't a full-time player, but I had a blast being on that team because we won the division and I was playing in the majors just 45 minutes away from my hometown," McClendon said. "Winning makes all the difference in the world, and even though the Giants beat us (four games to one) to go to the World Series, that was a great year."

McClendon's legend also reappeared in '89. Zimmer summoned him to his Wrigley Field office soon after he arrived from the minors and informed him that he was playing left field and batting fifth. In his first at-bat, McClendon sent a souvenir to the fans in the stands. "The club had been struggling, but I hit that home run, we won the game and we never looked back. We regained first place that day and won the division.

"The best part was that my family was there and they got to see me do something special on the major-league level. They knew, that day, that I finally had realized the dream that I'd been chasing since even before that Little League World Series in '71."

Despite the success, McClendon started the 1990 schedule back in Iowa, and although he was promoted near midseason, his playing time was more limited than ever before. After making 49 appearances, McClendon was traded to Pittsburgh in September.

"I was extremely happy about coming to the Pirates because Jim (Leyland) sat me down as soon as I got to Three Rivers and told me that I wasn't just there for the stretch run that season. He said I was part of the plan for the next few years," he said. "Then, in my first at-bat as a Pirate, I hit a home run.

"I wasn't on the playoff roster in 1990, but winning the division set a tone for the guys on that team. I think we all knew it was only a matter of time before the guys on that club went in different directions when it came time for free agency, so the next couple of years were going to be special."

Leyland used McClendon in both right and left field

and also at first base when Orlando Merced was rested during the 1991 season, and he responded by hitting .288 in 85 games. "He was a perfect fit for us," Leyland explained. "The one thing that's very important to a ballclub is knowing how to put a team together, and sometimes your extra player can be a very important piece – the right type of guy, the right type player. 'Mac' was the perfect fit because he was a threat to hit the ball out of the ballpark, he could catch, he could play first base, and he could play the outfield.

"He was the perfect piece to our puzzle, and he was also a very valuable National League player where you have double switches and things of that nature. He was a jack-of-all-trades and probably a master of none defensively, but he was very adequate no matter where you put him. I never worried about him or thought of him as a liability defensively."

The Pirates, stocked with the likes of Barry Bonds, Bobby Bonilla and Andy Van Slyke in the lineup and Cy Young winner Doug Drabek and 20-game victor John Smiley in the rotation, won the NL East again in 1991 with a 98-64 record. Pittsburgh, however, lost to Atlanta in the NLCS in seven games. After the season, Bonilla departed for the New York Mets to become baseball's highest-paid player at the time, and Smiley – who was due a $3.4 million salary – was traded to Minnesota for outfielder Midre Cummings and a 23-year-old southpaw named Denny Neagle.

"I think everyone knew Bobby was going to go for the money at the end of the '91, but when 'Smiles' was traded during spring training, we all knew this would be our last chance," McClendon remembered. "We knew Barry and Doug would leave, too, because of the situation the organization was in. They didn't have the money to keep those guys.

"But we played hard and we didn't give up," he continued. "And we ended that season with 96 wins and right back on top of the division. Then we went against Atlanta again, and I think every Pirates fan knows what happened then."

The Braves, with starter John Smoltz registering wins in games one and four, raced out to a 3-1 NLCS lead. Pittsburgh, however, outscored Atlanta, 20-5, in games five and six to force a Game Seven showdown at Fulton County Stadium. McClendon batted .727 with eight hits in 11 at-bats in the Series and successfully reached base in his final eight plate appearances with three walks, four singles and a home run in Pittsburgh's 13-4, Game Six, win.

Drabek took a 2-0 lead into the last inning of the final game, but following an error by Gold Glove second baseman Jose Lind and a single by Francisco Cabrera, former Pirate Sid Bream slid safely to complete a three-run comeback for the Braves.

"It's still painful to talk about that moment today," McClendon said. "I can close my eyes and see it. And I can feel it. That moment right there really took a lot out of me.

"For me, the next couple of years were half-way decent, but the mental focus, the concentration level and the desire was not there on the field anymore. I'll never forget the conversation I had with (Bill Virdon) in 1994. I said, 'I think I'm done. I can't focus anymore out there. I don't know if I want to do this anymore.' Virdon told me the same thing happened to him."

McClendon extended his professional career by 37 games after his final, 1994 season in Pittsburgh. He signed with Cleveland as a free agent and was assigned to Class AAA Buffalo to begin the '95 campaign. "Probably the biggest mistake I ever made in my life was letting the people in Cleveland talk me into playing again," he said. "I was there until the end of May, and even though I had decent numbers (.278 in 108 plate appearances), I didn't want to be playing anymore. I wanted to move on to the next stage in my career."

Former Pittsburgh general manager Cam Bonifay soon offered direction. Bonifay, the Pirates' GM from 1993 until his dismissal in 2001, contacted McClendon before he returned to his hometown and convinced him to visit Pittsburgh on his way from Buffalo. The results included a position as roving minor-league batting instructor in 1996 and placement on Gene Lamont's coaching staff as hitting coach for the 1997-2000 seasons. Bonifay orchestrated a five-year rebuilding plan under the new ownership of Kevin McClatchy, and despite owning the majors' lowest payroll for several years the ballclub batted a collective .261 under McClendon's tutelage. In 1999, Pittsburgh established a new team record for homers hit in a single season with 171.

Lamont's contract, however, was not renewed after his four seasons failed to produce an above .500 record. McClatchy and Bonifay interviewed 12 possible replacements, including Buck Showalter, Willie Randolph, Ken Macha and McClendon. On Oct. 23, 2000, McClendon was named the franchise's thirty-fifth manager at the age of 41.

"When the Pirates hired 'Mac,' and then asked me if I would be his bench coach, I said yes because of the respect he earned from me as a player," explained Bill Virdon, a major-league manager for 11 seasons with Pittsburgh, the Yankees, Houston and the Expos. "I think his intelligence is his best quality. I think he's sharp.

"And Lloyd respects the game of baseball with all his heart and soul, and that's something you don't find in players or managers these days. It's too much about the money – but not with Lloyd. To him, it's about winning, and if he's not winning, then it's about finding ways to win."

McClendon became the first minority manager in the history of the Pittsburgh franchise, but the hiring, he insisted during the press conference, was not based on the color of his skin. "I've said since the first day that I wanted to be judged by the content of my character just like Dr. Martin Luther King Jr.," said McClendon, who registered a 209-276 record after three years. "Can you be judged by the content of your character? I would hope so, and for the most part, I really have been.

"I've received hate mail and some people have said some nasty things, but it's all made me a stronger individual.

"And so has the criticism that comes with the job. When I sit back and watch football games, I yell at the television when I think a coach should have run a different play. But then I tell myself to shut up because I don't know why the coach made the decision he made. His quarterback could have the flu, or his best receiver could have turned his ankle walking out of the locker room."

Although Pittsburgh extended its consecutive-season losing streak to 11 campaigns in McClendon's third year as the Pirates' skipper, general manager Dave Littlefield extended his contract for a fourth season – a decision which met with Leyland's approval. Leyland, in fact, explained that McClendon's utility status as a player permitted his quick rise to major-league manager. "He played first base, so he knows something about the infield play. He also played the outfield, and he was a catcher, so he knows a lot about calling pitches in every situation," Leyland said.

"Hopefully, he's going to manage the Pirates for a long time. I think he's a good manager, and I think he's made tremendous strides. His club never quits no matter what the score is or what the team's record is, and that tells

Lloyd McClendon – 175

me a lot about his leadership."

"I think he relates to his players like I did," said Chuck Tanner, the Pirates' manager from 1977-1985. "You hug them when they need hugs and you kick them in the butt when they need kicked in the butt. At the end of the day, you've won as a team or lost as a team, and that's something that comes from the manager. I've seen that frame of mind from McClendon's club."

"People talk a lot about the chemistry in the clubhouse," Virdon added. "The media wants to know who the team leader is, and they act as if it has to be a player. Well, Lloyd McClendon has been the leader of his ballclubs, and they've played hard and they've played to win."

The Pirates opened PNC Park in 2001, and the rookie manager and his club were expected to win more than they lost because the new facility was to provide enough money for the organization to afford improved personnel. Instead, injuries and unrealized potential led to the termination of Bonifay and to the team's first 100-loss campaign since dropping 104 games in 1985.

McClendon promised Pittsburgh fans when hired that his primary objective involved returning the Pirates to the winning ways he experienced as a player in the early 1990s. But did he set his expectations too high when taking over a small-market ballclub at a time when disparity was a prevalent issue in the 30 cities owning major-league franchises?

And have McClendon's expectations been too high for him personally since being labeled with a legendary nickname? "As a child, I was very fortunate because I had parents and coaches who didn't place winning ahead of how you were supposed to conduct yourself as an individual and a responsible human being," he said. "Was I 'Legendary Lloyd' as a big-league player? Far from it, and I'll be the first to admit that – but I made it, and I'm not done.

"I want to win a World Series here in Pittsburgh. That's my one and only goal, and I'll be very disappointed if it doesn't happen. I know the business of baseball, and I know managers are hired to be fired. It was that way before me, and it will remain that way after me. But make no mistake – I'll win a World Series. I am a winner – always have been. I'll win."